Mythic Worlds
AND THE ONE YOU CAN
Believe In

BY HAROLD TOLIVER

Copyright © 2024 by Harold Toliver

ISBN: 978-1-77883-446-2 (Paperback)
978-1-77883-448-6 (Hardback)
978-1-77883-447-9 (E-book)

All rights reserved. No part of this publication may be reproduced, distributed, or transmitted in any form or by any means, including photocopying, recording, or other electronic or mechanical methods, without the prior written permission of the publisher, except in the case brief quotations embodied in critical reviews and other noncommercial uses permitted by copyright law.

The views expressed in this book are solely those of the author and do not necessarily reflect the views of the publisher, and the publisher hereby disclaims any responsibility for them.

BookSide Press
877-741-8091
www.booksidepress.com
orders@booksidepress.com

Contents

The Actual, the Hypothetical, and the Utterly False

Chapter One: Quarks and Made Up Things. 3
Chapter Two: Towers of Babble 25
Chapter Three: Quarter Moons and Fractional Truths . . 47

Why People Took To Collective Illusions

Chapter Four: Skull Duggery from Lucy to Mythology. . 73
Chapter Five: From Myth to Philosophy and Science . . 107
Chapter Six: Myth, Epic, and Oracular Sermon 139

Repair? Retrofit? Or Replace?

Chapter Seven: Modernism and Beyond. 171
Chapter Eight: Empiricism's Razing and Rebuilding . . 193

Myths of the Commonwealth

Chapter Nine: Militant Rhetoric and War Gods 217
Chapter Ten: Groupthink on Steroids 241
Chapter Eleven: Organized Turbulence 267
Chapter Twelve: Reining in Popular Illusions 301

References 323
Index 363

Epigraphs

The idols and false notions which are now in possession of the human understanding, and have taken deep root therein, not only so beset men's minds that truth can hardly find entrance, but even after entrance is obtained, they will again in the very instauration of the sciences meet and trouble us, unless men being forewarned of the danger fortify themselves as far as may be against their assaults. (Sir Francis Bacon, *The New Organon*, XXXVIII)

Unadulterated, unsweetened observations are what the real nature lover craves. No man can invent incidents and traits as interesting as the reality... The truth–how we do crave the truth! We cannot feed our minds on simulacra any more than we can our bodies... If you must counterfeit the truth, do it so deftly that we shall never detect you. But in natural history there is no need to counterfeit the truth; the reality always suffices.(John Burroughs, *John Burroughs' America*, 8-9)

I do not mean to suggest that the custom of lying has suffered any decay or interruption–no, for the Lie, as a Virtue, A Principle, is eternal; the Lie, as a recreation, a solace, a refuge in time of need, the fourth Grace, the tenth Muse, man's best and surest friend, is immortal, and cannot perish from the earth My complaint simply concerns the decay of the art of lying. No high-minded man, no man of right feeling, can contemplate the lumbering and slovenly lying of the present day without grieving to see a noble art so prostituted.(Mark Twain, "On the Decay of the Art of Lying")

The great enemy of the truth is very often not the lie -- deliberate, contrived and dishonest, but the myth, persistent, persuasive, and unrealistic. Belief in myths allows the comfort of opinion without the discomfort of thought. (John F. Kennedy)

PROLOGUE

People aren't as unique in fiddling with reality as they seemed before primate studies, but no creature else does innovation nearly as well, or wanders into error as often or as disastrously. Chimpanzees and monkeys can just barely scheme and deceive, and quadruped symbols of deception like the weasel and fox aren't in it. An adage for the public branch of concoctions came from Petronius in the first century, *mundus vult decipi, decipiature ergo*, "people want to be deceived, therefore deceive them." Long before then and ever since savvy ones in high places have found how well illusions serve their purposes in getting jobs done that take many hands. But presenting a public front isn't the half of it. We also endorse illusions personally and defend them with zeal. They range beyond politics in taking in the entire universe, but insofar as cults and other collectives gather in groups with influence they maintain a political side as well.

Our inventiveness includes innovation in weaponry, which makes armed illusions deadly. Where apes and monkeys managed no more than bites and scratches, the ingenious biped has come up with cudgels, slings, maces, bows and arrows, chariots, siege engines, and eventually canons, rifles, and atomic bombs, and for defense, shields, helmets, walled fortifications, and anti-missile missiles, none of them adequate. Much of the truth twisting has gone into gathering forces to use those devices. Preparation for doing so requires the invention of reasons. It is in the propaganda department that delusions come into play most aggressively, often among the ancients, and sometimes still, presuming no less than extraordinary high sanctions from the maker of the universe. That is where cults and nations often coincide. The biblical example joined Marduk, Mars, and others as a partisan war god that transmitted messages to the people through their patriarchy.

Such misconceptions made no demonstrable contact with the real universe, much of it brought within sight of the atmosphere-free Hubble telescope and of powerful electron microscopes capable of magnification up to about 10,000,000 times the size of focal objects. Compiling data and tested theory has put in a new light notions in the history of ideas that have lasted centuries, notably those concerning nature, cosmology, and the place of mankind in the scheme of things. Reclassifying the illusions as *poesis* would mean fewer crusades on their behalf, fewer fourth level jihad movements in the war minded branch of Islam, and no rival sect members and unbelievers tortured through all eternity. The problems of illusion-generated militancy would shrink appreciably if the basic dimensions and numbers of natural history were universally taught and what they say about untenable beliefs was brought into the debate. Science is often taught in specialty areas but seldom as an entire natural continuum.

That humankind is among the more self injuring species shows in accounts of lethal violence calculated in thousands of incidents, with cultural influence suggested by era differences, with higher rates between 700 to 1500 (120 per 1000) than currently (13 per 1000). Jose Maria Gomez in the Department of Ecology, University of Granada, extracts these and other figures from World Health Organization data and makes species comparisons decidedly not in the favor of mankind (Elaine Pagels, 2016). The toll from conflicts in the 20th century alone, put by some estimates at about 160 million, reiterates that appalling story. Many of the casualties we can attribute to errors in perception, to the propaganda that encouraged them, and to fervent beliefs contradicted by the natural continuum.

I won't be concerned with inbred brain modules or domains except to say that it seems doubtful that our vulnerability to delusions is due to any specific areas of brain architecture. What *is* unavoidable, however, is the conflict between ego or subjective point of view and objectivity. That is incurable for the obvious reason that any organism is first of all self oriented. We have to feed the singular mouth, see through personal eyes, and hear through personal ears. The brain is a very subjective instrument. At the same time any functioning organism has to live in a world not oriented around it. Getting these opposites in good balance is a lifelong task that requires constant adjustments as we learn more about what is out there. Casting ego and desire forth into what is truly there is a prime source of illusions. We humanize things easily, animating plants and animals, projecting human friendly creatures into sky and sea, some of them angelic and paying visits to our favorite cultural visionaries, some of them demonic. The most common corrective is familiarity with natural history. It sets the limits of what is real, and we needn't go far or look through magnifying devices to see enough of it to judge what is wildly improbable. Representative samples of natural history are all around us. What is far off and what is diminutive merely confirm its mixtures of order and disorder, beauty and irregularity, maternal kindness and predatory cruelty. The powers and dimensions are visible enough to teach us our proper place. Natural philosophy requires those who feel they have to animate a cosmic force behind all of that to make it accountable to natural history. When that is done the animation turns out to be indifferent to justice, prone to bring wreckage and chaos about, and responsible for animal suffering extending hundreds of millions of years.

When ideas expressed in words became possible, probably some 50 to 40 millennia ago, *Homo sapiens* gained in capacity to disseminate ideas in detail. That is what enabled groupthink and allowed the invention of fables. Discourse became the great enabler. It was free eventually to conjure angels from the clouds and devils from underground caverns as well as demons in enemy camps. Because it is just as willing to serve imagination as logic, distinguishing facts from myths became one of the more arduous and frequent things the brain is assigned to do. We have sung anthems, chiseled idealized icons into marble and put them into celluloid strips to simulate motion, but nothing works quite like speech and writing for delivering emotion and conviction in response to real and simulated things. Confused classifications are the chief means of *substitution* or *displacement* rhetoric, and of *vaulting* rhetoric, as when Philip II of Macedon told his subjects he was a god and they were obliged to agree, vaulting because the title temporarily promoted him from mortal to immortal. Demigods have been frequent in the annals of dictatorship and their myths, and a little of that added prowess descends into cult rituals, sacraments, and those who administer them. Where token evidence is offered for the claims, add evidence selection as a branch of part-for-whole substitutions. One aspect of something is used to characterize the rest. In the typical it actually does. In evidence filtering on behalf of bias it doesn't.

Through its 13.8 billion years the universe has shown no inclination to be at all like any of the myths of origin. The Hubble telescope's 20th anniversary image shows a mountain of dust and gas in the Carina Nebula that by itself makes any form of effective design hard to support. That's if jagged mountains and raging seas haven't already done so. The top of a three-light-year pillar of cool hydrogen is being worn away by the radiation of nearby stars, while stars within the pillar unleash jets of gas that stream from the peaks. The Eta Carinae Nebula, NGC 3372, is a gigantic miscellany of star clusters, dust, and gas trillions of miles in length. The shapes are only suggestive, the way a face appears in a rock formation or a cloud finger points across the sky. *Limbo* makes an apt spacetime metaphor for such a directionless, oddly shaped miscellany. All the galaxies put together appear to be a confusion, not a design, though galaxies do fall into clusters and superclusters. The human brain is itself something of a patchwork, albeit a marvelously functioning one at its best. The construction is like half planned architecture in which additions are tacked on as needed. Put nature's dimensions together by means of such a recording and categorizing instrument and again something like limbo fits the results, nothing definitive in either the processing or its dissemination. Nature may have its infernos, purgatories, and paradises, but overall it doesn't show directional movement other than the eventual expenditure of star fuel and what physicists characterize as total entropy, lacking in any further transfers of mass and energy under the famous equation $E = mc^2$.

In using such words as *truth* and *reality*, I'm making an assumption about objectivity that in the context of postmodern skepticism toward discourse needs justifying. One popular movement of the last half of the 20th century, in intellectual circles at least, distrusted whatever claimed to be unvarnished truth, thinking more of discourse than clever demons or of atoms. The number of references to Thomas Kuhn's *The Structure of Scientific Revolutions* (1962) testifies to that. That followed a long tradition of skepticism that hit a high (or low) mark in Descartes. I enjoy playing the 'see who can doubt the most game' but find that once you start on it, it consumes too much time and space. We can never be absolutely certain that what we think is real wasn't the doing of an extremely tricky demon, but it seems a pointless game. I'll just assume that the desk at which I sit is real, not an illusion made up of atoms and molecules. If the tricky demon proposal turns out to be true and we've all been fooled, we would be none the better for guessing that ahead of time. The pros and cons have been examined many times, recently and expertly by Sean Carroll in *The Big Picture* (2016). I don't consider *objectivity* tarnished beyond use. Nor do books like Peter Godfrey-Smith's (2003) that detail how science and philosophy collaborate in defining what is real. We rely on marked lanes of traffic despite an occasional driver who crosses the yellow line.

Nor does objectivity finish its work with what can be measured and described with certainty. Intimations in gestures and facial expressions are subject to more than one interpretation, but one version is usually nearer the mark than others. The A students in class are right more often than those who tweet through the lectures. Accuracy depends on facts and thus on objectivity as free of missteps as we can manage. Among the reasonably well established theories are the ages, distances, speeds, and numbers of natural history. Even the magnificent wreckage and rebuilding process of the Carina Nebula is relatively contained by comparison to the whole of what is visible. In total the cosmos may be unshapely but in physics, chemistry, and in contained systems like the solar system it follows invariables such as the speed of light in a vacuum. Until it is broken apart, every atom obeys a strong force binding its neutrons and protons together. It is also true that despite the overall movement toward total entropy, any given area can increase in energy and organized structure through its contacts with another areas.

In specialties remote from my literary background I rely on a good many scholars who have to be put together for a chronological picture of the natural continuum. These come to more sources than I like to inflict on readers, but both specialists and the rest of us have no choice but to consult studies remote from anything anyone can personally vouch for. For that matter, the universe itself has to be taken in samples that yield the invariables and constants. That is perfectly valid because what a handful of hydrogen atoms do in burning, splitting, and combining into helium has to be the same everywhere. With false starts and relapses, increasing the number of tested areas and combining them under comprehensive theories has been the general direction of intellectual history over the past several centuries, indeed from as early as ancient Egypt, Sumer, Babylon, and Greece. Natural philosophy depends not only on the sciences but on the humanities and arts. In that context *nature* includes information gathered from within the human sensory range as well as from methodical study. To adjust Einstein's saying: philosophy without science is lame, science without philosophy blind. (That's not quite what he said, but never mind.) Neither works as well without common experience as it does with it. The use of telescopes, spectroscopes, and microscopes depends on eyesight and its filtered passage into the brain's receptors. Everything is sized and its velocity gauged with reference to human proportions, never left completely behind even in numbers that reach into the dozens of zeroes.

I'm not concerned with everything that runs counter to natural history, merely our vulnerability to misconceptions that propose quite different universes in the interests of self identified ethnic groups and nations. Modern weapons of mass destruction under the direction of blind faith aren't to be taken lightly, nor is science denial that blocks efforts to avoid environmental deterioration and what may become quite damaging levels of global warming. That overlaps to some extent Noam Chomsky's account of media propaganda in *Necessary Illusions: Thought Control in Democratic Societies* (1989), but his concern is mostly US propaganda in political and foreign policy contexts. I assume that what is true of Americans isn't unique and that the roots of credulous belief are ancient.

Ordinary run-of-the-mill fraudulence I set aside together with individual machinations. Someone seeking to make an impression rehearses a persona offstage before presenting it at a board meeting or joint session of Congress. We expect that, and Erving Goffman (1959) has done an admirable job on "the presentation of self in everyday life." Whether or not that is a core self or a ghost in the machine is another topic I avoid. I also eliminate fraud and error limited to given disciplines. Medicine, for instance, has had its share of hoaxes and suppressed evidence. Quackery has gone public many times, as pseudo sciences such as alchemy and astrology once did and on a smaller scale still do. Going public in that sense doesn't raise armies, merely profits. Comprehensive accounts of the universe are what inspire collective beliefs and send armies forth. Because that level of illusion isn't based on reason and evidence, each variant tends to be hostile to the others.

That twisting the truth in modern scientific disciplines isn't more prominent is a tribute not only to those who go into them but also to systematic cross checking. Not much in the publishing arena is so thoroughly scrutinized as work submitted to professional journals and refereed books. In introducing *The Best American Science and Nature Writing* for 2006, Tim Folger remarks that proving a theory wrong is a favorite occupation in that shin-kicking industry. Despite that, histories of philosophy and science need chapters on errors, hoaxes, and partisan rejections of evidence. Theories eventually falsified have a purpose only if they elicit better support for better theories, as astrology and alchemy eventually did. "Something remarkable emerges from all the tumult," Folger concludes. "Even though the intellectual brawls never stop, charlatans are invariably exposed, and the ceaseless, collective, rigorous drive to find fault yields an understanding of reality impossible to achieve by any other means" (xii). *Invariably* is questionable, but the point is valid. We have good reason to trust methodical, peer reviewed findings more than we do most statements issued for public consumption.

Lest we condemn illusions and myths altogether we should remember that without them our distant ancestry might have remained too clan-oriented to build civilizations. Invention teamed with attention to detail has created everything that drives the streets and furnishes dwellings. It has manipulated the genes of domesticated creatures to bring them closer to what we want and of plants to increase yield. That is where Alfred Wallace, Charles Darwin and their forerunner the American William Wells in 1813 started, that is, with human rather than natural selection, the invention of dogs from wolves and of cows from wild aurochs.

Even us/them hostility based on conflicting world views isn't an unmixed evil, merely predominantly so. Some historians argue for armed conflict as the means of putting scattered provinces and city states together peacefully. A recent president of the United States attributed all progress to war. The Roman Lucan in *The Civil War* voiced similar ideas. Alexis de Tocqueville (2003), an astute critic of American history, thought the American colonies and states potentially chaotic without a strong federal government, and that government came about from the Revolutionary War and was maintained with civil war. Jefferson, Madison, and other founding fathers believed similarly. One military-minded student of civilizations, the Scotsman Adam Ferguson (1767, 1995), says with some credibility that the friendship, team cooperation, and courage of wars are ennobling. It is sometimes assumed that the best leadership arises in warfare. General Carl von Clausewitz as late as 1874 treated soldiers as pawns in the brilliant maneuvers of generals. That case for aggression as the drive train of progress is plausible only on the surface. The number of discoveries and inventions unrelated to conflict would fill a shelf of encyclopedias. Empires that fall within the range of archaeology and historiography–Sumerian, Anatolian, Hittite, Babylonian, Assyrian, Mittanian, Mycenaean, Chinese, Indian, Mongolian, Hebraic Egyptian, Persian, Greek, Roman–were brought together and fell apart by convincing themselves of things that weren't true. They talked and wrote themselves into magnificence and talked and wrote themselves into committing atrocities. Wars played a mixed role, stimulating invention and bringing ruin.

Believed myths include supportive or punishing gods and goddesses such as, anciently, Enlil, Shamash, Marduk, Re, Zeus, Jupiter, Yahweh, and currently a universal holy spirit. Stories of goblins, elves, fairies, nymphs, satyrs, the fates, gnomes, sprites, devils, angels, trolls, gremlins, vampires, and dragons are more obviously fabulous but less inclined to attract cults and lead to doctrinal conflict. In total numbers such projections of human psychology are nearly beyond cataloguing. Lists of ancient Mesopotamian figures alone if we include minor deities, spirits, and demigods numbered in three digits, some 16 of them major figures and about a hundred minor ones. Like ancient dynasties and like medieval and renaissance monarchies, both the Incas of South American and after them the Aztecs of Mesoamerica used such figures to rule. Law codes sponsored by divine counsels sometimes emphasized the protection of the weak and sometimes legitimized oppression. Among the Aztecs, myths conditioned people to believe that the functioning of the universe depended on tearing the hearts out of victims and holding them still beating up to the sun. The recipients of the sacrifices in Central America–Tezcatlipoca, Huehueteotl, and the bloodthirsty war god Huitzilopochtli–were intended to be intimidating, and so they were. Many beliefs and much doctrine seen in a social context are rhetorical enforcers. Their aim is to convince, not to explain how things work. That was clearly the case with the war god Yahweh and much of the time for Allah and God the Father adapted from him by Muslims and Christians.

The standard by which I gauge deflections from objectivity is equally lofty in the sense that it looks to natural history as the most substantial and well verified context of everything in existence. Given its debris, extremes, and cruelty, as I suggested and will repeat at critical junctures, I subtract animations from it to avoid unnecessary monstrosities. In contrast, notions of the cosmos that until recently prevailed nearly everywhere and still do many places were as distant from the truth as Plutarch's version of earth in likening its governance to the sun moving in "seasons in just proportions to the whole creation" (896). Much of what various populations believe is as far off the mark as that. The Jewish, Christian, and Muslim figures in the Non Sequitur cartoon are clutching thick tomes compiled before anyone knew the planet orbited the sun. They are dressed ornately and anciently. The equations on the board have to be worked around them lest they obliterate important portions of them.

We have come to understand the full scope of human and natural history only since radiometric rock dating, nuclear science in general, telescope-assisted astronomy, and advances in evolutionary biology. Astrophysics, geophysics, and chemistry underlie nearly everything including biology. As Hartmann and Miller point out in *The History of Earth* (1991), until the Dutch scientist Antonie van Leeuwenhock discovered single-cell life in 1677, lifeforms could not be studied "and classified according to their microscopic cellular structure" (101). Evolutionary biology awaited not only microscopes but the concept of natural selection's dependence on environment, which in turn derives from physics and chemistry. Astrophysics now goes to the beginning of those estimated 13.8 billion years ago. The story of life, earth branch, began over three billion years ago in single cell stromatolites and spent most of that span evolving into multicellular forms of life. What in the 17th century were separate studies of distant things seen in telescopes and minute things in microscopes are now connected in a single narrative. The discovery of subatomic particles, laws of thermodynamics, and mass/energy conversions was necessary to forge the links. What was once considered a great chain of being has become a chain of causes and effects. It can't be said too forcefully that this narrative replaces many another that prevailed almost uncontested until Darwin and what came together in the 1920s in the collaborations of geology, astronomy, evolutionary biology, and physics.

The most reliable data we take from three sources, sensory impressions, science, and chronicle history. *Poesis* adds a hypothetical or *what if* branch of learning based on simulation. I'll say more about that in the first chapter and at other points in defense of myths and fictions *recognized as such*. I've appreciated these enough to spend a career on them, including Milton's largely untenable version of the Jewish, Greek, and Christian versions of world history. We find added reason to value fictions in Rousseau's hating them as much as he hated science–and Athens and Catullus, Hobbes and Spinoza and China. As Sir Philip Sidney (1992) remarks in *The Defence of Poesy*, historians courted the muses and "usurped of poetry their passionate describing of passions, the many particularities of battles, which no man could affirm; or, if that be denied me, long orations put in the mouths of great kings and captains, which it is certain they never pronounced." Indeed "neither philosopher nor historiographer could at first have entered into the gates of popular judgements, if they had not taken a great passport of poetry" (214). The history/myth hybrids of the ancients en route to science and philosophy support that notion.

The initial task is to sort areas of discourse into the right bins, one of the largest of which mingles fact and fiction, sometimes meaningfully, sometimes confusedly. Quite a few lively and impassioned differences would melt away if the distinction between objectively validated truth and illusions was better recognized. How large is this subject? Obviously outsized. "To sort out… [the] philosophical issues of anthropology and archaeology is not only difficult, it is also boring on a scale imaginable only by people who have read the complete works of Hegel," Robert J. Wenke warns in *Patterns in Prehistory* (1979). Adding natural history, beautiful lies, misconceptions, and myths of state would be like throwing in the works of Gadamer. Hence I've settled for selected topics and historical samples that amount to reflections on myths and illusions. We can draw on methodical work without being methodical. Getting types of representation in mind with enough examples to indicate their uses advises taking core samples rather than attempting coverage. To test the climate of 100,000 years ago one need only sink a metal casing into ice and extract a sample, not excavate with bulldozers.

The goal we should work toward is the disassembling of harmful myths of the commonwealth and blind faith partisanship. What separates self identified groups and generates their mutual hostility is less important than the common human heritage. The myths block our vision of a reality that is far and away older, larger, and more formidable than any of the myths. That reality shrinks the human presence is to be expected. Reality checks often do. That is inherent in the maturation process from the infant's self-centered small world to ever expanding knowledge. That is perfectly acceptable. Ego should be dedicated to mind, not mind to ego. The universe is what it is. Having accepted it, we find it just as good to be a small part of so immense a reality as a larger part of misconceptions. We haven't actually ruined much of anything yet except areas on the outer rim of the planet that a diminishing number of species and our own progeny will inherit.

The Actual, the Hypothetical, and the Utterly False

Chapter One

Quarks and Made Up Things

Naturalist Numbers Don't Come Naturally

That life is not only hard at times but precarious is the main reason for imagining it improved. The home cures aren't working. The crops are failing. Enemies are at the gate. Maybe Marduk has an answer if we can get his attention. An imperialistic campaign dangerous to those set to undertake it could use a favorable sign from one of the war gods. Drawing on a gift to foresee the future an oracle supplies it. Meanwhile in stars numbered very approximately in the septillions, electrons, protons, and neutrons keep their appointments in numbers we would call infinite if the literal meaning of the word didn't place it beyond the highest number. It isn't the practice most places any longer as Sean Carroll (2016) points out to call what the atoms and stars are doing 'causal', but the mechanics do operate by invariable laws. It always takes 4 hydrogen protons plus neutrons, and gamma rays to produce a helium atom, throwing off neutrinos and positrons in the process. We aren't far off in saying that the heat plus the nature of the neutrons, protons, electrons, and the rest 'cause' the helium to form so long as nothing remains of intent in the word.

The entire natural continuum up to the point at which lifeforms start choosing within a narrow range of options is causal in that limited sense. So far as anyone knows nothing in the Periodic Table of Elements could have been any different than it is. Neither Marduk nor any of the other intangibles had anything to do with it, or with the tornado that touched down in Kansas today.

This first section consisting of three chapters applies the natural continuum as a litmus test to determine what is actual, what hypothetical, and what utterly false. The problem is always the same. We are capable of having quite firm convictions while mired in error and of feeling uncertain about what isn't all that doubtful. We can't change the brain's wiring, but we can improve the input and output. At its most comprehensive, the *actual* includes everything that exists of whatever size from quarks to quasars. About 16 constants are classified as universal plus half a dozen more limited to the electromagnetic force. We've no reason to question these or equivalents in the theorems of math. The *hypothetical* straddles the difference between what is possible and what is well established, usually, if we're being methodical, in a suspended state pending further investigation. The number of galaxies, their clusters and superclusters is uncertain at present but is getting better calculated and will have an improved tool for seeing into deep space when NASA gets its next telescope aloft. Axioms and theories are relatively assured but normally provisional because exceptions might exist somewhere as yet undiscovered. Something stated hypothetically in a formal way is set it up for critical scrutiny. The *utterly false* needs no introduction, only an explanation as to why we so often consider something real when everything is stacked against it. That Superman flies through the air despite being quite heavy and having no visible means of support isn't intended to be believed and so isn't false, merely fictional. That is the case for novels, plays, and poems but not for sincere beliefs such as those of *The Divine Comedy* and *Paradise Lost*.

What difference does it make, one might ask. Take one of the more prominent issues of the day. A stat sheet on terrorism compiled by thinkingbynumbers.org calculates that we spend 50,000 times more per death on terrorism than on any other cause. Over 30 times the 3,000 victims of 9/11 perish annually in hospitals inadvertently, which isn't to say don't go there if you need to. You are safer in Manhattan's One World Trade Center than in the nearby hospitals. Several thousand times greater casualties came in the world wars and wars in Korea, Vietnam, Afghanistan, and Iraq than in the 9/11 attack. These in turn continue what has gone on for thousands of years in the only species that has wounded or killed its own in the hundreds of millions. Thinking accurately is thinking more by the numbers in many cases. That is dull and kills the evening news, but if the goal is to arrive at the unvarnished truth we sometimes have no alternative. Connotations cling more easily to words than they do to numbers, though statistics, too, can lie.

We have a valid and unavoidable subjective reaction to everything we encounter and an equally valid objective one if we stop to think. Reasons for variability in perception aren't difficult to see. What is new or startling is more memorable than what happens all the time. News media looking for something to fill the time find more promise in what makes noise than in a hospital infection or heart attacks that afflict non celebrities. Moreover, defense spending is profitable, and exaggerated dangers can be used to generate it. Politicians hire spin doctors for similar reasons and invent slogans to substitute for meaningful discourse. Corporations hired researchers to show that DDT wasn't harmful. Fossil fuels have nothing to do with global warming their representatives in Congress maintain.

In daily life we sort things out on the run, usually making distinctions more or less reliably. No one could function very well without being right most of the time. As a discipline, naturalism teams up with philosophy to go a step further. Astrophysics starts at the beginning and proceeds with a mechanical sequence through to a projected total entropy in the range of a digit followed by over a hundred zeros in years. Where theories follow the Occam razor principle they eliminate anything unnecessarily elaborate.

As Newton rephrased that principle, "We are to admit no more causes of natural things than such as are both true and sufficient to explain their appearances." That's not a particularly debatable point, but which of two explanations is preferable often is. Does nature happen to follow invariable laws and constants, or is it designed to do so? The repercussions of the latter alternative aren't to be taken lightly. Nothing that intentionally made a universe mostly inhospitable to life, has caused nearly all species to go extinct on the one known habitable planet, and inflicted pain on the survivors is to be entertained indifferently. If such a power, intervening at will from beyond nature, isn't necessary to explain what exists we've no commanding reason to add it. Natural philosophy can make room for it only provided that the characteristics presumed for it don't contradict what it has produced. Adding a host of demons to explain what went wrong loses credibility when we consider that none ever shows itself. The size and age of the visible universe argue against it, as do such facts and figures as these: if only one in every hundred billion stars has an orbiting satellite that sustains cellular life, there would roughly speaking be a hundred billion such in the universe. A couple of missions in search of them going from here to the ends of the visible universe in opposite directions would search only narrow bands 93 billion light years each 5.88 trillion miles.

  Animations do have a place. We get slightly more familiar with the fourth planet from the sun (Mars) and even the far distant Pluto because of the Roman war god and the god of the underworld. The personifications don't share characteristics with the objects, but the names are easy to remember. Field, alpine, rosy, and wooly pussytoes–to go from the immense to the trivial–project kittens into plants that have almost nothing feline about them. Other kinds of feet were on the minds of those who named the partridgefoot and the coltsfoot. Animating things in that manner to make them more personable and memorable is subjectivity teaming up with science. Names like *beggarticks* and *sneezeweed* are catchier than *Bidens cernua* and *Helenium autumnale*. The lousewort has carried a tarnished reputation for centuries despite having never deposited a louse on anyone. That insertion of something familiar is useful and harmless and falls under convenience rather than falsehood. At least the lousewort does actually exist. *Ceres* never did. Groupthink being regional, we have less difficulty in seeing that seriously intended animations of other places are myths than we do with those in our immediate vicinity.

Nomenclature in General

Quite apart from mistaking parts for wholes and believing what isn't likely true, any attempt to sort out degrees of probability runs into an unavoidable mismatch between words and things, as pronounced at times as the one between geometry ('land measurement') and topography. Names are deceptive in the sense that they make categories seem as real as particulars. Not until we get from abstractions down to instances do names designate actual things. In saying *that white rose* we put the object in a botanical category, with *white* distinguish it from the kind's other colors, and with *that* designating something existing at the moment. The real thing. The generic category is in the hands of botanists and gardeners. Nature itself doesn't so much as add one thing to another to get two. It merely follows regular reproductive steps to produce more of the same, the seeds of things in Ovid's panoramic creation in *The Metaphorphoses*. When we say the *actual* it is *that* white rose we mean, not whiteness categorically or the *rose* as a kind. The universe is made up of constantly changing particulars. The mechanical procedures are the invariables and constants by which they come about. We can if we wish assign repeatable patterns a secondary existence but they aren't actualized until they appear in singular instances. That objects come in repeated patterns refers us to atomic, molecular, and genetic near replication not, as in Platonism, to pre-existing idealized forms cited by category titles. In a more obviously contrived manipulation, lions are one thing under naturalism and another in monarchical symbolism. In that kind of substitution of name for thing we have no trouble distinguishing between the symbol and the object, though some of the animal's strength and ferocity may carry over to a monarch in public perception. Only the man actually exists. The office he holds is a convention, an agreed upon artifice. Insofar as it alters his own brain work and puts a crown on his head it takes particularized existence and moves from an idealization to ontology. So do the works he accomplishes one by one, these too having to be particularized in instants to move from plan to reality.

Numbering is as conventional as the awarding of names. Calculations can map the patterns and chart the regularities, and to that extent we can endorse a mild anthropic principle and say that math too corresponds to constants like the speed of light, ever the same in a vacuum. It can specify the angle of a plane's departure from an assigned longitude and latitude at zero elevation. In fact math may be the only way to specify such things exactly. Newton's theory that gravity weakens by the square of the distance separating attracting masses is a confirmed calculation even though no one knows for sure yet just how gravity works, and as Ian Stewart (2016) points out, applying gravity to predict future motion quickly runs into figures that would disable a supercomputer (48-49), billions of trillions and trillions of billions. *Velocity* per se has no existence, nor does vector, angle, and length in the geometry branch of math. Only something moving at a speed from a starting point at an angle from another line has reality. Speed and vector and the position at a given instant are calculations of *it*, the objective thing. Thus a plane moves with respect to earth's surface along a line of flight calculated in terms of speed, longitude, latitude, and elevation.

As Scott Atran points out in *Cognitive Foundations of Natural History* (2004) and together with Douglas Medin in *Native Mind and the Cultural Construction of Nature* (2008), whereas the everyday names of folk usage work by readily visible likenesses and differences, what scientific taxa take into account often go unnoticed. Though quite different means of categorizing, these tend to be mutually supportive: "Folk biological groupings have always provided an intuitive underpinning and empirical approximation for the scientific species" (149). Noteworthy taxonomists such as Aristotle, Theophrastus, Dioscorides, Cesalpino, Linnaeus, and Darwin, have sometimes had to rearrange the categories of their predecessors to line up ancestry with offspring. Since Darwin and Wallace, the genetic heritage hasn't always been self evident. Although anyone can see that a dog has a different ancestry than a cat, not many would automatically link a Chihuahua to a gray wolf or a chickadee to a dinosaur. The history of a given Chihuahua has followed a DNA trail the individual instances of which were DNA carriers. The recurrence of the DNA combinations with slight variations, parting gradually from a line of wolves, can be said to have a formal or secondary existence, but again these too need individual instances to be actualized. Evidence for lines of descent is layered in strata and has to be sorted. Linking a sloth to ancient grazers of sea grass and seaweed is a relatively recent practice. That would have no bearing on the survival of legacy beliefs except for the illusion universal up to about Darwin that whatever has a name has existed since the hour of its making.

Common sense, science, and *poesis* have their own ways of handling resemblance and attaching connotations to semantics. Unlike science, which calibrates differences carefully, when poems put two things together one of them is normally more familiar than the other, and both express feelings that may be more important than the semantic citation. What counts isn't so much how well we know one or the other term but how far apart they are and what emotion flashes across: "My love is like a red, red rose" locates the less tangible love at a measured distance from the familiar red rose. We know the lady herself isn't literally in the rose family or symbolically thorny, that only the beauty of a very red rose transfers. As in gapping a spark plug, poets generally want just the distance that will make mental and emotional sparks fly but not so great as to seem absurd. My love is like a green, green onion would raise a chuckle rather than devotion to beauty. Of course anyone who actually said my love is like a red, red rose outside of a poem would already be suspect. Speaking in poetry is as unnatural as speaking in legalese or with the staged majesty of a leonine king.

When poetry revives worn-out metaphors, it does so by reestablishing a distance that has collapsed from overuse. The gap between the maker's hand and a tiger is the question of Blake's "The Tiger." What possible being could have made such a thing? The naturalist's answer is that the environment shapes every part of that and other predators, the claws, teeth, burning eye. The prey and the forest were already there, and over time the bright eye, fangs, and claws grew to take advantage of them, no shaping hand or eye involved. The connotations that come with the question drop the reader into depths of mystery based on the history of personifications. Blake's questions put these to a severe test. "Did he who made the lamb make thee?" is the key to that, implying that if you use the word *God* in the context of a tiger you've attributed to the maker something wild and cruel. A common theological answer would be *yes*, the divine maker made the lamb on which the tiger can feed if it chooses, and that makes the symmetry indeed fearful. "Did he smile his work to see?" The naturalist escapes the malice in that by assuming that the tiger just happened. Environment and biology did it, tough luck for the lamb, but that's the way natural heterogeneity works. Some things in an ecological combination collaborate, some devour, some get devoured. Ovid keeps enough ill tempered deities on hand to account for nearly anything including that ferocity, one advantage of a polytheistic set, though still not plausible for the billions of trillions scattered through spacetime.

In initiating a search for common ancestry and the rules of biological replication-divergence, resemblance follows another principle in naturalism free of intent. Nature never means to go from point A to point B. *How* it does so is the question, not *why*, although a naturalist might take up the latter question in order to dispose of it. At the level at which categories link instances, likeness and generalities belong more to taxonomy than to natural history, but the goal of science and philosophy alike is to get from instances to axioms or in this case laws of succession. These produce the likenesses and the category nomenclature, a filing and communication convenience. Making the connections is the joint business of observation and axiomatic philosophy. How stars burn and create elements couldn't be discovered without knowledge of subatomic particles. Poets, philosophers, scientists, and theologians alike were left without a blueprint until neurons, protons, and electrons were added to the record of names and concepts, the names being more or less arbitrary but having Latin and Greek roots.

Shortly after the major parts of atoms were discovered galaxies were identified and found to be moving away as if pushed from a condensed center. The master natural history narrative that took shape in outline in a few decades from the beginning to the middle of the 20th century, featuring first Ernst Rutherford's work with electrons and then astronomy and particle physics, dispelled a good many illusions that had fired the imaginations of differently oriented populations. The nuclear furnace we call the sun became something else entirely. Ancient civilizations might have fought over territory and other matters anyway, but adding illusions increased the intensity of their differences. None of this is to say that language, math, and reality don't come together nicely at times. Among fabricated buildup/breakdown/buildup objects, the Subject-Verb-Object sentence is among those capable of standing in for natural history's sequences. I'll reserve for later continental philosophy's and postmodernism's objections to discourse that simple and positive. The key to the sentence and strings of them built into paragraphs, monographs, and narrative structure is the active conjunction, the link between subject and object. Conjunctions work because both things and ideas have connections. Comment on that, too, I'll reserve for later, when the development of language can be considered as a landmark in the prehistory of *Homo sapiens*, indeed at about the demarcation line often set between *HS* and *HSS* some 50,000 years ago.

Disjunctions among things bring chaos and among ideas nonsense. Humans are unconnected to trees and rocks with respect to feelings but can be joined with them in touch or even something as ephemeral as light reflected from either and reaching the other. We have connections of all kinds with real objects that remain or keep moving when we are absent. I know the clock on the wall has continued to be there, because while I was gone it ticked off the same span as the wristwatch I compare with it. Someone might have tampered with it, true, but I was within sight of the only entry to the room and didn't see or hear anyone. I judge the odds against any outside interference to be negligible. The clock is real and my conjunction with it is easily renewed. S-V-O in the making: "I see the clock." No tricky demon allowed.

The Shape of Things

Another indication that something presumed to be historical has been doctored is strictly formal. *If a story has an announced beginning* ('I sing of arms and the man'), *a progressive logic, covers an extended span, and ends decisively, it is fictional.* The happy endings of novels, plays, and movies aren't the problem, but the fictions of traditional cosmology can be. If they displace a great deal of natural history they become supreme fictions. Neither natural nor human history unfolds in a decisive way over long spans. It is from the present that the past is reconstructed and a conclusive future is projected. The present can in fact indicate what the past had to have been to bring it about, and the future can be predicted, but enhanced history and anything comprehensive about the future are a different matter. It is present misogyny that created Eve and Pandora and present misery that creates a future without misery.

On the extended time scale of evolution the impulse to move everything including species toward betterment did what it could with the mechanics in the later 19th century, when evolution was sometimes made out to be teleological. Had it not extracted humans from high primates? That was crucial to the effort to salvage providence and intelligent design when survival of the fittest argued otherwise. Ruskin was among the first to head in that direction. Without going to his extreme, Stuart Kauffman (1995) proposes an overall evolutionary advance from bacteria to humans that makes the latter at home in a universe graced "with a bounty of order." Meanwhile we spin "around an average star at the edge of a humdrum galaxy" (Kauffman, 71). Biological organisms can't be projected into the macrocosm with anything resembling comfort. Cellular life doesn't venture abroad physically without a closed environment equipped with oxygen, water, and provisions, and venturing abroad mentally likewise runs into the discomfort of extreme temperatures and distances outside everyone's comfort zone. However adventurous and ingenious future inventors get, humans are never going to colonize the universe. The realistic guidelines stories that work under different laws than the known ones are set for us by natural history. That a long range narrative ending decisively is fictional, however, doesn't hold for segments singled out and assigned boundaries for story telling purposes, as a biography can use birth and death for natural boundaries. The Olympics have opening and closing ceremonies. Terms of office are set to the hour. Someone could make up a plausible life story of a molecule, as Gamow's Mr. Tompkins imagines the life of electrons. Entire civilizations begin and end, though in that case with vital components before and after that lead another life. Some natural beginnings and ends fall within our sensory range, but enlarging the framework weakens the relation of parts. Anything inclusive is also inconclusive. The miscellany factor and the huge numbers in the microcosm and macrocosm are disorderly in total if not in all local areas. If the big bang gives way to a big crunch, making cosmic expansion and contraction into a pulsating cycle will produce an endless story. That cosmologies and world histories before the 20th century were fundamentally flawed is revealed as much by their decisive beginnings and endings as by their miracles. One day the creation got underway with a command, and another day as yet unspecified it will come to an end by the same means.

Before the radiometric dating of rock, spectrometers, and powerful microscopes and telescopes, both cosmology and earth history were based on limited observation. Not until Alexander von Humboldt's *Kosmos* (1845, 1858) was an extended and at times well theorized earth history put together. Up to that point it generally fell about four billion years short of the mark and cosmic history over three times that. Impressions strictly from within the sensory range can be misleading. A star close to earth looks larger and brighter than an equivalent farther off and neither it nor the rest orbits the planet as it appears. Even before the mind starts its processing of sensations the senses have already been at work on them. The brain filters what has made it through, sorting, putting similar things together, and composing schemes to account for them. Telescopes and microscopes, too, were used at first to emphasize wonders rather than the fuller story they would eventually help piece together.

The fleas and flies that Royal Society members Henry Power (1623-1668) and Robert Hooke (1635-1702) watched under magnification impressed them with the symmetry and beauty of the eyes and the agility and strength of the bodies. Isolated from other facts, that was surprising. Awareness of marvels caused the brilliant Hooke, master of half a dozen disciplines or what later became disciplines to conclude that "we shall in all things find that Nature does not only work Mechanically but by such excellent and most compendious, as well as stupendious contrivances, that it were impossible for all the reason in the world to find out any contrivance to do the same thing that should have more convenient properties. And can any be so sottish as to think all those things the productions of chance? Certainly, either their Ratiocination must be extremely depraved or they did never attentively consider and contemplate the Works of the Al-mighty" (Vickers, ed., 1987, 128). Thus the quandary of Blake's 'what if', which Hooke like nearly everyone of his times failed to think through. He needed only to imagine the tiger, lamb, and smile together to see the problem, or if that didn't work, include an earthquake, a drought, or a plague. Lacking a concept of millions of years of nature-modified lifeforms, most of them failing at some point, Hooke's conclusion seemed logical within the constrictions of the prevailing world history. Even using common observation, however, he was ignoring the fragility of the insects and their mortality. It wasn't unknown that they came forth, grew, died, and disintegrated.

The information that instruments added wasn't really needed to show other irregularities in flora, fauna, and geography. If geometry sets a standard for symmetry, much of nature is irregular. It has few if any parabolas let alone anything as nifty as a hyperbola. The planet's tectonic plates move unevenly, sticking and then lurching, leaving cracks and raising mountains–earth's broken crust stood on end. Both the irregularity and the catastrophes were as evident in the 17th century as in any other time. Shorelines are rough hewn. Peaks, valleys, canyons, swamps, and forests are unshapely and home to ecological mixes whose components both collaborate and fight for space. Sir James Jeans (1932) is understating the case in saying that the universe "appears to be actively hostile to life like our own" (4). The habitable margin is extremely narrow: "At a rough computation, these zones within which life is possible, all added together constitute less than a thousand million millionth part of the whole of space" (6), an estimate on the generous side since the visible universe continues to deepen and what is visible is only a small part of the whole. He wasn't of course allowing for the possibility of billions of other habitable places.

Unlike science and natural philosophy in accounting for nature's mix of order and confusion, works of art subordinate details to a design, and it is just such crafted artifacts that set a model for lofty concepts of the universe. They are part of our projecting into something else what is a human construction. Coherence through and through is a sure sign of intelligent engineering. Art can be complex but isn't allowed to be random. Henry James in "The Art of Fiction" (2004) finds it "a kind of huge spider-web of the finest silken threads suspended in the chamber of consciousness, and catching every air-born particle in its tissue" (434). That stresses sensitivity and perception more than order, but it also underscores design. It is the structure of the web that holds it in place and makes it receptive to whatever floats or flies near. In brief a poet reconciles things that in the real world are at odds. W. H. Auden (1962) calls an accomplished verbal artwork a community of such reconciled feelings and substances, but a verbal one only and warns against thinking that because "all is well in the work of art, all is [also] well in history.... all is not well there" (71). Actually all isn't well in art either except in the sense that it is coherent. Thatched huts in the forest have witches living in them who bake and eat little children.

Consciousness we hold responsible not only for imaginative invention but for value systems. The latter make use of nature but stand apart from it as framed art does. Codes of conduct may take natural history under consideration, but they are set by common agreement and framed in prescriptive language. Making a human cease to be is normally condemned except when nations call for doing it on a large scale. (The 'thou shalt not kill' tenet of the decalogue is violated wholesale at the command of the figure who presumably issued it.) Justice, compassion, temperance, beauty, courage, glory, generosity, foolishness, and sadism are sufficiently removed from anything material to justify Thomas Huxley's separation (in *Science and Morals*) of mental things from external things. Comparatives such as *more glorious, wiser, kinder*, and *crueler* are judgmental. They assume an average compiled from numerous instances that in most cases haven't been methodically tabulated. They are constituted both *within* and *against* nature.

Bacon's distinction between history and *poesis* offers an explanation not for value systems per se but for hypothetical and imaginative projections of value. Made in the divine image in his view, the soul has a higher sense of perfection than anything it finds in nature. Though that doesn't account for monstrosity myths, guileful figures like Isis, or treacherous ones like Marduk and Orestes, it does suggest why images of perfection are appealing and why we like to think them real: "The use of this Feigned History hath been to give some shadow of satisfaction to the mind of man… wherein the nature of things doeth deny it" (186), hence golden age myths. Ending a fable with poetic justice remedies a glaring defect in history and is one of the appeals of fiction. The reason for heroic modes is that people seek more ample greatness and more exact goodness than history normally illustrates. They desire magnitude, and so "poesy feigneth acts and events greater and more heroical." History, Bacon adds in *The Advancement of Learning*, lacks what we crave. The episodes of history "poesy endueth… with more rareness, and more unexpected and alternative variations" (1996, 186) than they originally had. In *The New Organon* (XLV, 1985) he adds that: "the human understanding is of its own nature prone to suppose the existence of more order and regularity in the world than it finds." "More heroic" fits epic. "More golden" fits pastoral. More orderly and regular fits the idealizing imagination of the visionary poet. What poets mistaken to be prophets imagine they and others hold to be real. That is a simple case of getting the genre wrong. The principle reasons for doing so are contained in Bacon's statement. Nature and life leave us dissatisfied and we want "some shadow of satisfaction" they don't provide. Making life less hard is the reason for most invention, making clothes, building shelters, taming fire, inventing the wheel, and projecting fictions as if they were real or at least possible. The adoption of *suffer* (to bear with) to ordinary events like voting rights derives from the sense of patiently waiting them out, or impatiently not doing so as in 'he doesn't suffer fools gladly' and the Suffrage movement of the 1920s.

Sidney, too, in *The Defence of Poesy* (1992) finds poetry infiltrating history and philosophy, "For whatever the philosopher

saith should be done, he [the poet] giveth a perfect picture of it."
The main connection between the imagined worlds of *poesis* and
the legislative schemes of morality is the use of the former to
urge the observance of the latter. In that office, fictions become
partly rhetorical. Sidney isn't fully justified in saying that the poet
"nothing affirms, and therefore never lieth" (235). Even in openly
acknowledged fiction, artifice does affirm and does lie. Parables
and moral fables indoctrinate. Because fables illustrate precepts,
he is, however, correct to find them livelier than cut-and-dried
philosophical axioms and better at moral instruction than chronicles
confined to what happens. Despite inevitable biases, poems remain
basically detached and ironic in the sense that what they say is not
often what they mean. The play *Coriolanus* pretends that Coriolanus
walks the stage when it is an impersonator who does so. The moral
of the story is the moral of the *story*. Even when poems and plays
are guided by theses that readers and audiences carry away with
them, they lend themselves only secondarily to partisan counsel. The
Divine Comedy, however, expects its readers to believe in the universe
it presents.

    Poets mistaken for prophets are in the forefront of moral
legislation and likely to be preserved in that role from generation to
generation. From their positions as spokesmen for invisible powers
they acquire considerable leverage in legislating behavior, foremost
loyalty to a given cult. Faithful adherence to a set of beliefs and its
administration is the prime requirement for making a predicted
happy ending come true. The Babylonian *Epic of Creation, Erra and
Ishum*, and the incantations of the Hebrew prophets seek to convert
imagination into reality by means of appeals to capable powers,
especially in bringing curses down on enemies. Erra's curses have
a Hebraic sound in vowing to destroy the rays of the sun, cover the
moon, stifle rain and snow, ruin cities, bring down mountains, and
kill all cattle and wild beasts (Dalley, 297). Except in the presumed
influence of a powerful force, that resembles an epic catalogue
that Homer, Virgil, or Statius might have offered. A listing that

includes all ranks of society prevents any segment from escaping. To avoid damage an entire population must be found acceptable to a priesthood authorized to represent the power that inflicts the damage and confers the blessings.

That the supernatural authority a prophet brings down on lax and idolatrous populations is a device converts what looks like a curse into an exhortation with moral aims, in other words into rhetoric. No deity is actually going to bring Assyrian troops into Israel to punish delinquents in the faith, but Amos believes that will happen and hopes to influence behavior accordingly. Natural processes are responsible for disease and mortality, but assigning the power to inflict them to an animation gives its representatives awesome authority. Incantation and ritual that from one standpoint are merely words thrown into a vacuum are presented as a functional address given magical powers. The speaker doesn't acknowledge his role playing, because any hint that the prophet is actually a poet would convert the high tone into parody. Nearly everyone disbelieves in the seers, sibyls, prophets, angels, demons, and divinities of other times and places. Only something handed down through a local credentialed tradition is excepted. Whether the coordination that results is applied to good or bad ends, it is undeniably effective.

The somewhat bedraggled term *beautiful* applies almost equally well to fabulous narratives, doctrine-guided fictions, mathematical formulations, and selected aspects of nature, some of them as grand as stars seen from a distance. Certainly beauty should not be limited to the Greek *aesthetikos* or the Hellenic emphasis on symmetry, balance, and just proportion. As Ellen Dissanayake points out in *Homo Aestheticus* (1995), it comes in many forms and combines with other elements in aboriginal, Chinese, and Japanese art. Before those forms angled off into cultural variants, they would have followed an instinctive love of coherence universal enough to be called inborn. Disorder and confusion are unpleasant. No creature likes them. Art has controlled amounts of them and nature more than effective design would allow. As far back as we wish to go, the animal heritage must have profited from clear and distinct identifications entering episodic memory and preferred it to doubt. A bear rising up on its hind legs, sniffing, listening, looking, is seeking to identify something. So is a staring pronghorn and a watch-guard prairie dog. An often mistaken or confused member of any species is less likely to survive. That is only a probability, however, not a universal rule. In nature's miscellany with many things going hither and yon, faring well is also a matter of blind luck. Randomness and rule come paired in situations not governed exclusively by invariables and constants.

Imagination and Reality

We can translate the fundamental distinction between artful contrivance and factual accounts of natural and human history into simpler terms by borrowing Wallace Stevens' interactive imagination and reality, on occasion altering it to mind and object or subjectivity and objectivity. In Stevens' case the interaction is applicable strictly to *poesis*. Both that and natural history look for patterned order but one is free to manipulate the objects the other seeks to account for. The miscellany nature has produced shows by contrast the orderly schemes that categorizing imposes on it, in science with a common consensus nomenclature. Some debris and litter are left unsorted. Go moderately small a moment and consider a few square yards of ground. A beetle, say a twelve spotted Lady Bug, moving unevenly along is an organized creature. Not much within its range is related directly to it except by proximity. A fallen leaf obstructs its passage, and it moves around it. If it had the concept and the language, it would classify such obstructions as random litter. It is also moving in concert with the planet as a whole without anything in sight showing that. We need Einstein to make that principle into special relativity. Its movement at the planet level is highly regulated and is patterned yearly to the second. The sun and chemistry are essential parts of our hypothetical sample, including the soil and the minerals underfoot. These belong to mineralogy until plants make use of them and turn them over to botany. Thus in terms of design or order, some miscellany, some uniformity, and constant change in everything except the fixed positions of things moving at the same rate in the same direction. Nothing is not moving. Enlarging the scene to the solar system shows the planet's movement with respect to Mars. The Milky Way moves with respect to Andromeda. Every integral thing is its own rigid object and therefore a frame of reference. Its molecular and atomic movements are enclosed in the single frame.

As with a beetle on its patch of ground so with everything moving everywhere. At the cosmology level the number of moving things is beyond telling and the vectors in total are a confusion. In the pliable and viscous bodies of Einstein's general relativity (1916) the integral objects no longer have to be rigid, and since size doesn't matter we might just as well go down to the particles that make up protons, neutrons, and electrons moving with respect to one another. Toss in photons, neutrinos, and gamma rays while we are at it. The combinations quickly overwhelm anyone's capacity to keep track. Order, however, can usually be salvaged in terms of the placement of contained moving objects, as many heated molecules colliding in a glass tank are disorderly but the tank itself remains shapely and stable. Geometry and calculus can be applied to it. Randomly scattered grazing cattle aren't an integral object. Panicked in a stampede they momentarily move in the same direction at the same speed, becoming more or less one. Nothing can be considered fully answerable to natural history and natural philosophy that doesn't take coordinating and differentiation principles into account. The relations of everything in existence depend on movement, but coordination has an exception to moving in the same direction at the same speed, namely uniform rhythmic succession. Repetition can be patterned like surf coming ashore. No object can be related to a fixed source by either movement or rhythm, because no unmoved center exists. It was for contrast to transient things that Aristotle posited an Unmoved Mover, a durable point of reference in much theology and philosophy for about three millennia until everything including species was found to be in motion.

In a version of the unmoved mover influenced by Cambridge Platonism, Marvell's "On a Drop of Dew" offers a narrative of what the poet considers a typical soul's travels from and back to a permanent source. Let us take that as an example counter to the beetle encountering litter. The poet offers no reason for the descent from that fixed place to begin with, an omission in Plato as well, but the descent puts an unchanging essence into a world of motion and into peril among transient objects. The soul's condition as a dew drop resembles what Kant in *The Critique of Judgement* (1790) calls the *subjective aesthetic*, "free from confusion either with concepts of the Object or sensations as determining grounds" (147). Once its momentary embodiment evaporates and it returns to its source, it apparently won't be representable. Nor is the unmoved source itself. Because the soul no longer has movement relative to the sun it will merge into it. The poem's metaphors–the dew drop, the fountain, the flower, and the sun–have unused physical characteristics meant to be discarded as everything material is. One difficulty with that concept among others is the relation of intellect to soul and body. Tangible sensory life among real objects is primary to a mind's processing of experience where reality is never capitalized or set apart. The only concession Marvell's soul makes to living experience is its insecure trembling in the flower standing in for all material reality. It doesn't learn anything or process sensations into precepts, changing episodic memory, as some brain studies put it, into semantic memory, one primarily sensory, the other primarily verbal and mental.

A similar cleanup of undesirable factors comes with other idealizations and utopian constructions, but whereas a traditional utopia concentrates on just a polity and its social order leaving the planet as it is, the dew-drop soul's celestial paradise disposes of everything. Material things vanish. The sun itself becomes a bodiless divinity of pure light. Only geometric roundness is real, and it too is inherent in the *Almighty*. That radical disposing of everything in the interests of an unmoved mover is an extreme version of discarding evidence selectively.

The Naturalist Essay and Rhetoric

In its most rudimentary form the naturalist essay is a step-by-step account of an experiment. It adds very little that is artful in rendering a sequence that reveals a pattern. The first naturalist essays with a scientific interest rendered just such accounts of observations and thereby became a branch of natural philosophy. Except travel journals that cited new lands and strange cultures, nothing of quite that kind existed in English before the Royal Society in the 1660s. In England if not always in Montaigne essays hadn't been narratives in a setting but definitions and collections of related apothegms. Thomas Sprat's history of the society, written before it had much history (1667), champions experimental science. Fact-based essays and narratives of sequences came hand in hand with the specimen collections of land and sea voyages. The latter would eventually send Darwin to South America, Franklin north, Shackleton south, and Livingstone into Africa. A combination of adventure narrative and verbal mapping, the naturalist essay developed into a hybrid form that besides narrating and cataloguing could also seek to persuade. In commenting that one who walks finds "his faculties unsheathed, his mind plastic, his body toughened, his heart light, his soul dilated; while those cramped and distorted members in the calf and kid are the unfortunate wretches doomed to carriages and cushions" (39), John Burroughs (1875, 2001), writing on "Exhilarations of the Road," exploits the essay's openness to championing untamed nature. That became one of natural philosophy's claims to priority over other branches of essay literature. Along with its empiricism came a defense of earth itself. It was something to be endured in Plato and in other worldy theology.

The lecturing tone is less casual when Aldo Leopold (1935, 1991) remarks concerning a groomed forest in Germany, "I never realized before that the melodies of nature are music only when played against the undertones of evolutionary history." In the German forest that inspired the *Erlkonig* "one now hears only a single dismal fugue out of the timeless reaches of the carboniferous" (229). That we should be hearing more derives from a commitment to biodiversity missing in row-planted trees. "The timeless reaches of the carboniferous" takes in a typically lengthy span in the naturalist essay, and the "dismal fugue" reminds us how often civilization's assaults on the planet throw it into mourning.

Chapter Two

Towers of Babble

A Reliability Index for Discourse?

Wouldn't it be nice, one thinks, if noses really did grow longer at least temporarily with each fabrication put forth as real. As it is we have only common experience and science to separate reality from semblance. I propose in this chapter to test key discourse kinds from casual to methodical to see how well they match up with what geology, physics, chemistry, astronomy, and common experience say about natural history. *Can* we sort out chronically misleading illusions on that basis? I believe we can to a limited extent by learning to be more habitually objective, though that eventually gets us involved in how the brain processes sensory experience and invents what it finds wanting.

The more elaborate illusions varying by area and era probably developed alongside the mind's entertaining hypothetical propositions. Some of the most implausible ones became the most fervently believed. Getting up in arms over them has been a chronic human failing, so we know cures don't come easy. If the universe we have in sight or have otherwise detected is real, certain alternatives proposed for it can't be. As we saw last chapter, no means of representation matches up perfectly with reality, and so right off we have a problem with disseminating and teaching natural history and applying it to natural philosophy. Math is limited to what is structured in a calculable way. In geographical matters it fares best with geometric flat planes, spheres, and rhythmic waves, the latter measurable by interval, height, and velocity.

Most math calculations work by balancing quantities on both sides of an equal sign (=), or unbalancing them on both sides of an unequal sign (≠). Or indicating an approximation (≤, ≥). Such formulations have limited application to irregularity. Calculus can get laborious in just putting an arc into an equation, and nothing in either nomenclature or numbers does well with scattered and shapeless confusion. Its density, boundaries, and any uniformity among its particles are subject to description but not the collective movement and collisions. Photography captures confusion in its raw state and in motion pictures something of the turmoil.

Mechanical computation gets further than mental computation because machines work faster and more accurately than brains and have exact recall. The speed, volume, and accuracy of artificial intelligence are impressive, and the help that gives its users is particularly worth noting because it has had some success in charting partial confusion. The history of how that came about may not have originated with the Turing machines of War II, but it might as well have, mechanical computation saw such advances from the 1940s to the end of the century. Roger Penrose's (1989) account of that history and questioning of whether or not consciousness is more than machine-assisted calculations (1994) concentrates mainly on the overlap of the computer and the brain. Working free of computational error has proved to be achievable in several areas but not in quantum mechanics or dark matter and dark energy. That the latter exist can be seen in their effects, but not being made of atoms dark matter doesn't give off information directly. In quantum mechanics, neither point particles nor waves make an adequate figure for what B. K.

Ridley (1984) describes as a swarm dancing attendance on electrons (118). Such things as vacuum fluctuations and "half quanta" can't be accurately gauged, because "we have to use one bit of the Universe to measure another." The bit a physicist uses influences the bit being measured (109). For particles combined in atoms, however, measurements can proceed as if quantum uncertainty didn't exist. They are locked into the isotopes and elements of the periodic table and can be observed without disturbing them. The metals and other elements on the table are well known by atomic weight as are the weights and structural strength of alloys. Skyscrapers, bridges, and planes are the result. They require only reasonable inference, not the deductive certainty of theorems.

Even within the sensory range the depiction problems continue. Topography, for instance, is too irregular and confused to be put in exact figures and words. Catchall descriptions like *rugged* have to serve. We have no problem with generalizing nomenclature like *random* or *chaotic* but a good deal with more exact representation. Chaos theory and Mandelbrot sets have gotten results with pictorial variants that find some structure in complex and oddly shaped fields. Beyond a certain point, however, intelligibility fails as does the anthropic principle. Much of the universe is fashioned so as to be understood and formulated but not all of it. Fortunately, concern with intractable illusions and misconceptions can set aside some of the limits to knowledge, and study can turn to what is patterned without numbering the leaves of the trees. Many of the quicksand situations are well known. As John D. Barrow (2000) points out echoing Bacon, extrapolating too much order from nature in disregard of the asymmetries and irregularities "is an understandable by-product of an activity that sets out to codify and organize our knowledge of the world" (332). As a consequence, physicists, Barrow tells us, "have made progress as much by overthrowing bogus laws of Nature as by discovering new ones" (332). That can be said of other areas as well. Progress in the history of ideas consists largely of replacing impossible or unlikely propositions with better substantiated ones.

The rules change completely with articles of blind faith not subject to evidence. Once one of them has become habitual, its proponents have an investment in it. It wasn't just the church that resisted Copernicus and Galileo and clung to the geocentric solar system. A good many others did as well. No one who professes disbelief in the conventional myths will be elected president of the US in the foreseeable future. Rituals meant to influence invisible beings remain widely practiced, the beings differing by region and era. The Iban headhunters of Borneo, looking to recharge their sagging fortunes by collecting newly severed Chinese heads, might have noticed sooner than they did that their expeditions and ceremonies were having no effect. Their barbarity was a semi-ritualistic form of discourse using acts rather than, or along with, words, much like ancient sacrifices accompanied by raised arms and chants. Whether or not actual toads get thrown into cauldrons, equivalents to "Double, double, toil and trouble" are still being tried many places to see if they might not work *this* time. Chants and rhyme, however, are now more likely to be parodied than taken seriously. (In one staging of Dvorak's *Rusalka* a witch throws stuffed animals in the caldron.) Hag speech and rhythmic incantations that summon a Beelzebub or Mephistopheles, dark versions of orthodox rituals, aren't in use any longer except in imaginative fictions, though magical transubstantiation still converts ordinary substances into powerful ones. Despite all the rituals, prayers, and pleas practiced daily around the world, species continue to go extinct, the environment grows less sustainable, faulty hearts continue to fail, and good fortune and misfortune maintain their customary ratios. So far as we know, nothing human or divine has ever altered the planet's orbit or spin or prevented a single atom outside the solar system from the fission/fusion in store for it under a certain heat.

Anything habitual such as appeals to magical cures indicates a cast of mind, which is what we would expect from what Francis Bacon assigns to different sides of consciousness (Vickers, 1996): "we raise our Imagination above Reason; which is the cause why

Religion sought ever access to the mind by similitudes, types, parables, visions, dreams" (218). What counts in a rational explanation is the observed object and its likeness to similar objects. In the quest for reliable axioms, a single fact contrary to a proposition unseats it. Thinking about what Karl Popper and Paul Feyerabend say to that end, Lee Smolin (2007) finds imaginative science working through an initial stage in which it cites provocative but incomplete evidence. The computations may be suggestive but at that stage are incomplete, like those of James Bradley in 1728 and Leon Foucault in 1862 that came close to gauging the speed of light. When tests are then conducted, the hope is to have something beyond reasonable doubt, which doesn't mean absolute certainty, merely higher probability. An international consensus can then cut down on entrenched beliefs to the contrary. Should a hypothesis fail even one test it must be amended or discarded. High myths without any such validation survive by virtue of traditions not subject to testing. They don't easily cross regional or cultural boundaries because indoctrination in them remains mainly local. Whatever evidence they produce isn't convincing outside a given circle.

Keeping the leverage evidence provides in mind, let us sample kinds of discourse, beginning with the kind we use the most and know the best.

Casual Talk

Two scenes of mixed information and error are the dining table and living room. When we sort through kinds of representation we haven't finished with social tossed salads until we include their casual talk, which shouldn't be underestimated in the spread of information and means of indoctrination. Outside of outright lies, the variable forms and moods of gossip are the least reputable, most casual, way of expressing opinions and rendering impressions. We don't look to them for theories of gravity, but their influence in religion, politics, and even in science and philosophy is considerable. Casual talk characteristically mixes opinion, information, hearsay, hyperbole, humor and some purposeful misinformation. We can readily see why it is culpable in spreading misconceptions. It is a valuable extension, too, of friendship and courtship and the most common means by which personas go public. Training of the young and lasting agreements get their start from it. In lowering something overly exalted, it sets the record straight, drawing on mockery and parody, the folksy way of deflating something.

That puts it among the more influential forms of exchange, competing with the podium and pulpit in disseminating public opinion. Customs are consigned to its care, as one imagines Iban elders over dinner teaching their offspring about the potent magic of severed heads. Most of the early language training of children comes casually, the influence passing from older to younger family members. The Roman historian Polybius goes so far as to say that historians should circulate among their contemporaries to gather information by word of mouth. That is more purposeful than ordinary gossip and more professionally sifted, but it is true that personal contact elicits more details surrounding events than memorial prose, diaries, memoirs, letters, or news, or at least details of a different kind. That courtrooms probing for truth draw on the personal narratives of eye witnesses reminds us that gossip, too, conveys eye witness reports. '*I* saw, *I* heard, *I* felt' precede the testimonials. That opens up avenues to bias and casual information in the company of likes and dislikes.

Consider an exchange in Iain Pears' *The Dream of Scipio* (2002) in which Claude Bronsen, a Jewish businessman born in Germany, plays host to friends in the midst of War II. Bronsen traces the course that each item on the table has taken, foie gras from Dordogne; Dover sole from the Atlantic; Vandée lamb; cheeses from shepherds and from farmers with goats, sheep, and cows; wine harvested and processed "with a dash of inspiration"; cigars from Havana. That network of regions and occupations comes to bear on conversation with an international bent:

> Gossip? You say. Idle chitchat? Yes, gentlemen. Men in trenches, men starving, men in chains, do not have the leisure to gossip. Gossip is the product of spare time, of surplus and of comfort. Gossip is the creation of civilization, and the product of friendship. For when my friend here made his inquiry he passed on the information necessary to keep the delicate fabric of friendship together. A question about a friend known for decades but hardly seen, an acquaintanceship which would fall into the past unless its shade was sustained by the occasional offering of gossip. And think again: My friend, an Alsatian businessman, was asking a question of a half-Italian writer about the marriage of a Norman lawyer and a Parisian lady of faintly aristocratic origin. All this at a dinner given by myself, born a Jew. What better distillation of civilization is there than that? ... Repeated often enough [gossip] binds society together (141).

Distillation is the key. Each guest takes in not only the products of shepherds, fishermen, farmers, merchants, and wine makers but events filtered through successive minds. That is as much a process of reasonable inference as methodical science is. It merely omits the quantifying, as estimates normally do. Hanging over the pleasantry of the dinner is the worst devastation of the three eras of the novel, classical, medieval, and modern. Bronsen finds civilization closing down despite having been "the finest product of the mind of man." It is closing down because of rabid beliefs spread by Third Reich propaganda, a top down distortion pretending evidence of a conspiracy it has invented. The guests will soon turn into beasts or victims, repeating age-old patterns as gossip commonly does. Barbarians, as the Scottish historian Adam Ferguson once pointed out, build their hovels in the shadow of palatial ruins. Having misjudged the strength of the French army and the swiftness of the German advance, Bronsen himself is overtaken, arrested, and imprisoned. Within a few months he is dead of pneumonia and malnutrition, a bitter irony after a feast based on cooperative trade and friendship. Illusions captured, imprisoned, and killed him, illusions magnified by becoming Third Reich policy.

What Bronsen doesn't say is that gossip helps spread the hostility racing across Europe. It is partly by word of mouth that the network of fascist operatives dooms its victims. The roundup of Jews would not work, at least not as effectively, but for this individual eyeing that one and whispering. We are reminded that in social novels like Jane Austen's household talk is the main engine of the plot. Before the advent of electronic devices it was person to person talk, podium and pulpit, speech, essay, and treatise that planted ideas and reinforced customs. Whatever stature the influential have is propped up by it and now its equally prolific and unreliable electronic forms.

Systematic Nomenclature

Thomas Huxley (1893, 2004) following Darwin was among the first to put incessant change in the prehistoric context of evolution and make a philosophical point of it. Nothing until evolution had quite matched Copernican and Galilean displacement of the planet from the center of the universe. Huxley's observation that all things are "the transitory forms of… parcels of cosmic substance" wasn't in itself new but including species was in Wallace and Darwin. People could now be linked systematically to the universe in more substantial and detailed ways than in any of the prevailing myths. Environment governs evolution, and the physics and chemistry of the universe govern environment. Besides being made of atoms and molecules, mankind evolved in response to the composites these formed in oxygenated air, water, land, plants, and animals. No one in Huxley's day or among the Greek atomists knew what the indivisible building blocks of matter were. The atomists for their part knew only that a minimal unit had to exist if motion was possible. Otherwise, every space would be infinitely divisible and impossible to cross. A matching paradox possibly from Zeno's teacher Parmenides applied to time. Because no moment intervenes in which things could move, every given moment must find things in the same place as the preceding and following moments. To Huxley, particles had wended "the road of evolution from nebulous potentiality, through endless growths of sun and planet and satellite; through all varieties of matter" (50). *Nebulous* is justified by the large scale heterogeneity nature produced by minute increments. The future stages of evolving forms became unpredictable when so many of them were in the mix and the changes came in such small fractions.

Systematic nomenclature begins with the smallest units and builds piece by piece, natural law by natural law, up to the largest visible configurations in galaxy clusters. It is ruled by cause and effect in all disciplines. It also concerns incessant change, but no language or math can handle that in detail. That the periodic table is an index of molecular units makes it one of the most reliable forms of discourse. After disintegrating, the elements of a compound move on to something else. Finding anything that doesn't break apart and combine in something different carries down to subatomic particles, waves, strings, or fields, the title *atom* ('uncuttable') again having been used prematurely. The breakdown as well as the buildup is the material basis of systematic nomenclature. Though the smallest possible things don't divide, some of them collide in swarms, combine, attract, repulse, and annihilate one another.

Social aggregates present special difficulties in naming and measuring because their individual units are highly complex and add volition to mechanical and biological means of buildup and breakdown. As David Hackett Fischer (1970) points out, they aren't closely analogous to any material body: "A human group is something more than a heap of people and something other than an organism or a machine or a great person or an idea. A group is not exactly born, and it does not precisely die. It has a beginning and an end, but no life cycle, no organic pattern of growth and decay. It has no roots or branches; no fruits or flowers; no mind or heart or soul; no cogs or gears or wheels or levers. It does not possess a will or a personality" (216). We have trouble doing without such analogies between social orders and other things, and trouble with them. The comparisons explain and misconstrue simultaneously. People are wolfish sometimes but never wolves, have pecking orders but aren't birds.

Ordinarily we have no trouble removing what doesn't carry over in an analogy, as in discarding the thorns, roots, and pedals of a red, red rose. An overall social body as opposed to more condensed suborders within it isn't integral and doesn't move as a whole relative to other earth-surface objects, only together with them and the planet. It is an amorphous, ever-changing abstraction. As much a concept as a definable thing it sort of exists as a real thing and sort of doesn't. It is like a bag in which to put groceries, except not like a bag either, since it has permeable borders through which its members come and go. Its networks are like fish netting except not regularly configured. Its lines of communication come and go by the moment and so aren't a labyrinth. In naming such aggregate societies we aren't using the same logic we apply to a class of vertebrates or an integral material body. They have sub groups that are cohesive enough to move in unison. Even when individual members of such groups are strangers, they can do so, like passengers on a plane. That makes them a systematic cohort that can be accounted for in a manifest or cargo inventory. Very few group movements are more systematic than dance troupes, marching formations, and embarking and disembarking passengers. In their assigned seats, members who are dispersed and separate elsewhere take on the movement of the inclusive body. Internal groupings based on ideas and beliefs are especially amorphous.

Given the complexity of its individual members, the internal movements of a human group are too complex to be charted in particulars. To establish relations where they are tenuous or non existent, political and legal discourse is littered with fictions of convenience such as that a trust exists on the same footing as those who established it or that "all men are created equal" when they are equal only before the law. C*reated* is a fictionalized version of 'born' assuming a supernatural hand in a biological mechanism. Social discourse is full of such half fictions and conveniences. Some of the wider gaps between names and things we find in the stereotyping and hasty labeling that come with ideology and doctrine. They assume the uniformity of designated groups composed of diverse members. Such group labels can be advisable at times on the grounds that indecision can be worse than error, and the details in any case would be overwhelming. Tribal members spotting intruders in a rain forest must decide immediately whether to advance or retreat. Best to assume they have poison blow darts and intend harm. That way a mistake won't be fatal. Armies wear uniforms and the Taliban beards for similar reasons. Even so, the uniform and the flash impression are at best partial stories. They seize upon one factor and make it stand for the whole, a problem with most category labeling that looks more systematic than it is.

Systematic in other words isn't synonymous with *accurate*. Doctrine and ideology can be systematically wrong. The nomenclature pyramid is misleading if it implies a reality for upper tier names equal or superior to that of instances, down where the details are innumerable and the combinations ever changing. Obviously the higher the classification the more detail it sifts out. Pearness exists only in individual pears, not a startling insight but counter to essentialist thinking that grants transcendent reality to nomenclature categories. Except for the inconvenience, we could omit *pear* altogether and go directly to *fruit*. Individual instances are seldom exactly alike, merely enough so to justify their being lumped together. The referential value of many words is arbitrary to begin with. Since we've come so far from the Old English *buttorfleoge*, it would be almost as appropriate to speak of smarms of flutterbyes as swarms of butterflies. Adverbs and adjectives fare no better than nominal classifications in catching some part of the whole and ignoring the rest.

As with words, so with math calculations. Some of them match phenomena well enough, but nature isn't obligated to provide perfect circles to match the unit circle with the vertical Y axis and horizontal X axis. It comes close in the forming of molten spheres, but currents, a roiled surface, gamma ray eruptions, and streams of photons and neutrinos disturb them. Gravity's capacity to make geometric shapes out of movable materials depends on their flexibility. Molten masses become spherical until they cool, as they do when erupted from a volcano and flow down hill hardening as they go. The result is a broken field. That makes a common mixture of the systematic (the volcanic cone) and scrambled detail. The spherical planet with its topographical features is the largest example within the reach of travel. It is regular enough to make navigation exact and grids logical from degrees down to minutes and seconds and yet is alive with movements and wrinkled with clefts and crags. Even a generally flat ocean surface has waves and currents and a mountainous floor like the land. Air swirls in numerous eddies and moves in jet currents and cyclones that keep meteorologists busy tracking.

The more movement and the more wrinkles, the less systematic the nomenclature can be. The visuals of a televised weather report are elaborate with the details of topography, temperature and wind gradations, cloud cover, shifting hours of daylight and dark, and tide charts. They are intricately beautiful and at the same time symbolic abstractions with only token likeness to the real things. The labeled pictorial grids atop land and sea and within the air envelop are good examples of system imposed on virtual chaos, omitting the clutter of cities, highways, moving aircraft, sea craft, and land craft, landfills, and scattered litter. Some of these down to merchandise and wrecking yards are subject to inventories where the real clutter comes in signs in columns and model numbers. Clay tablet inscriptions 5000 years ago were the first attempts to systematize miscellany, including in that case king and god lists. The gods, goddesses, and other spirits were litter cast forth from the symbol user to typify natural phenomena.

Natural History

The naturalist essay first gained recognition in the 17th century in Isaac Walton writing on fishing (*The Compleat Angler*, 1653). As it became more observant it added the presentation of experiments in the Royal Society mentioned earlier. Later naturalist essays filled the encyclopedias of the French compilers Compte de Buffon and Georges Cuvier before expanding into full-length studies by Jean-Baptiste Lamarck (1744-1829) in biology, Erasmus Darwin (1731-1802) in botany, and Charles Lyell (1797-1875) in geology. It followed from there into a post Darwin flood of essays on flora, fauna, and topography. Finding a place for them in natural history required their incorporation into deep-field time and species history. I choose this genre for its variability and capaciousness, for its plunge into the forests and seas of natural heterogeneity, and for its awareness of nature's rugged harshness. It characteristically mixes systematic nomenclature and observations in the rough in the common terms of the walker, climber, and diver. It doesn't trace any great number of contacts through a mixed ecology, but it does name and describe what an observer can identify in given scenes, often in revealing detail.

Reading history backward as the source of what is visible is a common procedure in both science and the naturalist essay and results in one of the more reliable forms of naturalist discourse. Rocks in that context become as much the ruins of past geological events as fossils are of life. When in the opening chapters of *The Mountains of California* (1894) John Muir makes western glaciers the product of "snow flowers," glaciers become to storms what blooms are to water, sun, soil, and plant genetics, namely, evidence of shaping events governed by natural processes. The metaphor fills out the cyclical mixture of growth and recycled materials with an encompassing mountain-altering sweep of time. The delicacy and beauty of individual snow flakes is ironic in the collective power they exercise as ice in partnership with gravity and friction. Glacial erosion, canyons, and moraine deposits are their products. We know how the compaction works once they have landed and fused into ice. What plate tectonics and volcanoes have raised, wind, ice, and water reduce. Farms and crops come in gritty downslope valleys. We can reason backward from a loaf of bread as far in the contributing chain as we wish to go.

In Muir that kind of time-span language is attention-getting in its own right, but when he finds himself splayed out on a Mt. Ritter cliff with an abyss beneath him, as a creature of rock, gravity, and fragile biology he too becomes an illustration. When the present meets the ancient, the match up is terrifying: "After gaining a point about halfway to the top, I was suddenly brought to a dead stop, with arms outspread, clinging close to the face of the rock, unable to move hand or foot either up or down. My doom appeared fixed. I *must* fall. There would be a moment of bewilderment, and then a lifeless rumble down the… precipice to the glacier below" (64). The word *doom* is distinct from the fated casualty of Homer, Virgil, Lucan, and Polybius where gods set the terms. In this context, *fate, destiny*, and *doom* mean what gravity, rock, and biology dictate, the first (gravity) there forever, the second in a cyclical phase that began with the cooking up of minerals, and the third emergent over a long span. As Muir's sense of gravity and elevation grows acute, the incident becomes a naturalist's *exemplum*. Inner-earth heat, tectonic plate shift and uplift, and glacial erosion have set the scene. Life fragile in its biology is on trial against an implacable mass. The moment becomes a frozen instant of a mountain-making phase and of an evolved biped whose hands and feet, are barely adequate.

The biological narrative is what produced the musculature, which reflects the chronic pull of gravity over formative ages. The brain's directing of the will reflects the survival advantage of courage and ingenuity. Knowing that won't help Muir out of his predicament, but it will help him write the essay if he lives. When he forces himself to resume climbing, every obtruding rock and recess gets his full attention as a handhold or foothold, fitted, or not fitted, to the bipedal form. That part is both chosen and random. Together environment and biology, brain and body, constitute a moment-by-moment struggle converted into a synopsis in the essay, properly so of course since there's no need to reconstruct the universe to portray the dilemma. Easier partnerships of ecological cooperation resume down in less barren terrain, but as retrieved in memory the incident reminds us that living conditions are narrow in range.

How narrow and fragile they are is a recurrent refrain of naturalist essays. Even those expert at naming and describing what is there can fall prey to misinterpretation, as Marston Bates does in *The Nature of Natural History* (1950) in going in a few pages from calling "this competition, this struggle" a superficial thing, "superimposed on an essential mutual dependence" (108), to accounts of parasitism and the competition that results from an individual fungus producing 700 billion spores, a tobacco plant 360,000 seeds, a salmon 28 million eggs in a season, and an oyster 114 million eggs in a spawning (172). As Malthus and Darwin demonstrate at length, struggles to survive aren't superficial but close to the heart of evolutionary history.

Muir moves on from what becomes a successful climb to take in more of the Sierras, still a harsh environment but less drastically so. The point of view often narrows and broadens like that, up close, then out and away. "Prospect" as a theme carries over, first the prospect of his surviving or not, then the more extended mountain range. Plotting an intersection with what comes next is essential to survival, more important than the past to the pragmatist Thomas Dewey (1917), because the moving event challenges us to meet what is coming: "the finished and done with is of import as affecting the future, not on its own account Anticipation is therefore more primary than recollection; projection than summoning the past; the prospective than the retrospective" (14). In that respect, what matters isn't what formed the mountains and developed grasping fingers out of ancestral fins and quadruped claws but what can now be seized to move the next bit up or down.

Adjusting to perspective shift and applying the past to the next encounter is a marvelous talent that goes unnoticed most of the time. It is what enables tying the here and now to various summoned spans of time or levels of categorizing. That same flexibility can lift the self to any level of objectivity the mind wants or needs for the next encounter. *I* can be placed instantly anywhere on the objective map and with practice can adjust feeling to the occasion, as Muir does in calling on will and courage. The journals of explorers like Shackleton, Livingstone, and Franklin likewise pause to survey. The broadening in this case puts the Sierras together in perspective: "Here are the roots of all the life of the valleys, and here more simply than elsewhere is the eternal flux of nature manifested. Ice changing to water, lakes to meadows, and mountains to plains. And while we thus contemplate Nature's methods of landscape creation, and, reading the records she has carved on the rocks, reconstruct, however imperfectly, the landscapes of the past, we also learn that as these we now behold have succeeded those of the pre-glacial age, so they in turn are withering and vanishing to be succeeded by others yet unborn" (70).

The accelerated motion of *withering* and *vanishing* puts hyperbole to apt use. If normal human measurements are to be engaged where geologic time is the subject, some such variability in the time frame is inevitable. Zooming in and out, focusing close up and far away, and linking the various perspectives is parallel to the handling of the verbal ladder from instances to categories and linking here and now to what comes next. Subjective and objective points of view never come separately. They are always *I* and a context selected by time and circumstance from the immense total. Comparatives are as common to adjectival and adverbial qualifications as they are to measurements. When April becomes the *cruelest* month, the language places the other months lower on the cruelty chart. Topography and earth history are comparative from one region to another within the normal sensory range.

The cultural frame of reference for earlier essays in natural history was broadly speaking European. For four centuries, European travelers going westward included fortune hunters, immigrants, militant invaders, refugees, runaways, and explorers, each carrying a point of view into the prospects of the new world. We don't know what the North American wilderness was to the first migrants from Asia, but to those coming from cultivated lands it was overwhelming, and they described it with amazement. Probably so much that was formerly unknown to migrants never struck so many as during that prolonged movement westward, which began in the late 15th century and was still going in the 20th. It transformed European perspectives on the planet. That the immigrants devoted themselves as much to conquest as to inquiry brings social and economic history into the naturalist essay and cultural difference into social relativity. A good deal of information reached Europe in inventoried accounts of Native American culture, topography, climate, and geography. It came with shortcuts and stereotypes. In more scientific terminology the term 'savages', for instance, was relative. Native Americans had an advanced non-savage culture compared to Australopithecus. Later Native Americans traveling by jet would be quite advanced compared to Europeans traveling to North America by ship or across it on horseback.

Another side of migration and perspective adjustment didn't become clear until later. Donald Culross Peattie comments unhappily concerning the exploitation of resources that "we have wasted, we have robbed and slaughtered and made wanton ruin of our wealth. History convicts us of setting fire to our forests, the last great stand of hardwoods in the world, because that was easier than cutting them down. Much of our incomparable system of lakes, brooks, mighty rivers, we turned into sewers where no fish but the worthless German carp will live. Our marshes, cradle of a million water fowl, we drained for crop land we did not need" (155). That was a different kind of inventory focused more on breakdown than buildup. Among the casualties are billions of passenger pigeons along with the millions of buffalo and thousands of forested miles that had initially astounded European immigrants. In the "veracious recordings" of the immigrants in essays, books, and personal journals we glimpse "deer, elk, antelope and bear, raccoon and fox, water fowl and salmon, whose profusion at the time of the white man's coming made this virgin land the richest in wildlife he had known within the memory of his race" (154). Looking back with regret is the mind's elegiac side in which loss becomes not an objective inventory but personal. The naturalist's long range perspective on the planet turns plangent frequently, especially where species extinctions have no possible recovery.

Amateur naturalist observation was also devoted to identifying the flora and fauna of the New World, usually under folk classifications rather than Latin titles. What John Burroughs (in *The Art of Seeing Things*, 1904, 2001) sees and hears in a winter wren is typical of earlier naturalists where they were realist observers rather than transcendentalists. That the bird has flown into a forested temperate zone links the specimen to an ecology. "Such a dapper, fidgety, gesticulating, bobbing-up-and-down-and-out-and- in little bird, and yet full of such sweet, wild melody!" (252-253). As an environment, the hemlock woods tell of climate belts, an ice shelf long retreated, and sun power transformed into biomass. The solar system in motion is implicit in every "fidgety, gesticulating" motion of the subject. So is perturbation descending from far off into the twitches of small muscles hooked up to a brain on guard for good reason. No way existed to be very reliable in such observations until the more extended planet history could be sketched in as background. With birds that longer range came to include dinosaurs and an asteroid collision random in the intersecting courses of two large masses. The ongoing encounters of differently sized and oriented lifeforms underlie the nervous watchfulness of Burrough's bird and many another that pecks and watches, turns its head quickly from side to side, and pecks again, foraging into the future.

Bunching Birds

I have taken liberties with Burroughs' twitching bird to point up one way naturalist narratives work. Bird are useful as samples of that because they are familiar to everyone and yet are often strange and exotic. Wonder is a subjective factor that in our sighting of birds comes winging in from nearby and distant hidden places. That makes them both common and elusive at the same time, more familiar and more strange the closer the imaging moves in. They also offer themselves to symbolic representation and prototypes, or I should say we seize upon them for that purpose in our habitual manufacture of symbols. Mythic creatures such as Mercury, who can fly with a couple of small wings attached to a helmet and heels, Hermes, griffons, devils, angels, and gods capitalize on winged flight. Birds appear to have been their models, as they are in cartoons for the newly saved, wearing small angel wings and haloes while sitting weightless on clouds. Some birds serve doubly as real creatures and symbols like the US eagle and the albatross of Coleridge's ancient mariner, the first not mysterious, the second opening upon a Coleridgean museum of horrors in the plotted encounters of fantastic narrative. It becomes the object of naturalist attention in the nature essay and fictional naturalists like Patrick O'Brian's Dr. Maturin, an exemplary scholar of flora and fauna in species-rich locations around the world.

  The evolutionary trail that produced actual feathered creatures hasn't proved easy to trace, but the evolutionary trail has become well marked. The resemblance of red crossbills to other Fringillidae, for instance, is something any bird watcher can see, but an evolutionary ornithologist might wonder how something as odd as a twisted beak came about when it looks dysfunctional for any but a specialized food supply. There couldn't have been many in-between forms of coniferous cones suitable for in-between beaks. Passeriformes (perching birds) emerged in the Paleogene with archaeopteryx in the far distance, possibly adding feathers and flight to hopping dinosaurs. Those stages are imaginable, but the crossed bill or scissor beak is unusual enough to suggest something odd about the environmental prompts. Does Darwin's gradual descent-with-modification or does the punctuated equilibrium of Stephen Jay Gould and Niles Eldredge better account for such an odd tool attachment? Oddity in the thing observed forces opening the book of nature to its systematic biology and evolutionary depths, as it is often the curious sight that stimulates investigation, theory, and perhaps eventually natural philosophy if the curious fact links up to something broadly hypothetical.

The function of the crossed bill once it evolved isn't in question, and like many naturalist observations it provides on going evidence of how evolution works. When BBC's *Planet Earth* used slow-motion to catch it levering open the cone of an evergreen and extracting a seed, the principle of specialized body mechanics matching an environmental niche was self evident. That falls in line easily with the natural selection theory. Where a niche exists, something that can exploit it will likely evolve selected from what works best. One remaining question is whether beak development jumped at the opportunity like the beaks of Galapagos finches at changes in rainfall or whether in-between states found in-between food that rewarded gradual development. That is a matter of the tempo of change, not of its probability, but linking change to time is a vital matter in putting together earth history. Evidence from several fields combined sets the beginning at about 4.5 billion years and situates the planet not only in the solar system history but in the astrophysics of supernova life spans and disintegration, the process that throws off masses of material to be gathered into lesser masses by gravity.

If the answer about the red crossbill isn't already out there in this case, it may be eventually, but following a fossil trail is one of the more arduous tasks scientists undertake. I haven't run across anyone who follows this one. We do know that scissor bills are occasional in other birds including chickens but are malformations due probably to the wrong position of the hatchling in the egg. No such malformation can lead to a genetic change. Craig Benkman, a red crossbill scholar, does find variants adapted to particular cones, which argues for a history of gradual adaptations. How bodily features function isn't usually puzzling even if exactly how they got that way might be. Another example: in equatorial steaminess, coming in for a landing as if out of a dream, a male Williamson's sapsucker lands on a deck railing. His keen eyes are set off by white streaks above and below. He looks very much all there, head erect, eyes focused. Finding nothing of value in a non native geranium, he takes off in a feathered rush, stretched out like a racer, beak and tail extended. Every bit of him—weight, bone density, speed, bill, focus—functions in the transfer of morsel-fueled energy to flight. That time-and-circumstance engineered form works as efficiently as attachments on a hand vacuum. It looks designed but developed over an extended period that wasn't kind to interim experiments. The intensity and the nervousness of birds announce *veracity,* as icy winds do to the Forest of Arden's exiled Duke Senior, and the veracity applies to anthropology as much as it does to zoology. Colorful migrants fly within range like confirmations, vouching for climate and geography.

Quickness and alertness need no argument as useful mechanisms, and streamlined motion is one of nature's strong points, balanced against weight and strength. The arms and fingers of the mountaineer clinging to a rock face were developed in the trees. The legs grew longer on the savannah. The brain expanded through ages of watching and learning. Like the bird in flight, he survives at the edge of his habitable zone, an extremely narrow one occupying a small percentage of the weights, temperatures, and element combinations at large in the universe. The marvels of efficiency that Power and Hooke admired in fleas and flies are evident in things that obviously don't need a microscope to be seen. Animals that flee do so at a pace encouraged by predators. Animals that choose to fight do so with requisite bulk, sharp tools, and strength. Animals that use deceit do so with camouflage and caution. Victorians who were dismayed at how their ape ancestry looked compared to them should have been impressed by ape intelligence, strength, and dexterity, the ape's millions of years of survival, and the additional millions of years of adaptations that produced the fingers and the brain. The journey from forest to savannah took courage and tenacity.

Once naturalism passes from current species to their evolutionary trail, which contains no splashes of color or bird calls, it enters less accessible but still partly verifiable territory. Where ephemeral traits aren't part of the recoverable record, inference fills in, another perfectly valid part of the systematic procedure behind the formulation of a hypothesis on the way to a theory. Paleontology has learned to be content with bones of the past and chatoyant flashes of the present. The scrutiny of facts and international naming agreed upon in the nomenclature link the essay to earth sciences and move by degrees from casual observation through meticulous recording to the general principles of evolution. Picking the right feature for a species name is somewhat incidental among the details. The flight that puts the sharp beak in position, the claws that cling to the tree, and the keen eyes that guide the sapsucker to the spot are functional parts assembled without a blueprint, but with the meeting of what the biology needs with the sap the tree needs intercepted on its course to the needles.

Among the many kinds of discourse, systematic nomenclature based on observation is the most internationally recognized and free of misconceptions. We rely on that as we do on unequivocal air tower instructions, longitude and latitude, and stop signs. The native babble of tongues carries its users in many directions, some of them colliding, but in naturalism it picks a wave length and a nomenclature agreeable to most who tune in. Acknowledging the common heritage in that regard sends people forward in roughly the same direction, which puts them on courses with fewer collisions than do perversely selected alternatives to what is as demonstrably real as crossed bills and intent, streamlined flight.

Chapter Three
Quarter Moons and Fractional Truths

Ritual and Algorithm

Strange pairing it may seem, algorithm and ritual, and indeed they are mainly opposites, but they have in common something that can be taught as procedure. An algorithm guides investigation where research has to be staged in the right way to produce results. Ritual approaches its invisible power through formulaic movements and words. The difference is equally as obvious: algorithms either pan out or not but in any case are applied to the quantities and numbers of material nature. Rituals that appeal to intangible powers have no demonstrable effect. The accepted method in both is usually ancestral or at least well known, though that hasn't always been the case. Scientific methodology got something of a fresh start when Galileo applied it to the conservation of energy and to bodies however bulky or feathery falling at the same rate (in a vacuum). It was by means of procedural ingenuity and testing that he found energy or force to apply not to velocity per se but to acceleration. As to algorithm proper, he believed nature to be written in math and used that supposition to calculate trajectory and other dynamics of motion. That set the scene for Newton and the development of the partnership between physics and math.

One of the growing realizations of renewed learning in the Renaissance was that if such a thing as a still center or unmoved mover existed it was unapproachable. By definition it couldn't register clearly or very fully in natural phenomena. In that respect, it is absolutely true that absolutes have no place and cannot be understood, represented, or subjected to method, hence the use of ritual with respect to supreme beings in their several versions. Nothing that moves is absolute, and all things move. The trick in that apparent contradiction is that the first *absolute* refers to discourse and the second to actual placement in the cosmos. The absolute to be thus denied a place would, if it existed, have unlimited power (omnipotence) and occupy infinite space (omnipresence) and be an unmoved mover. It was also usually thought to be all knowing (omniscient) so that nothing lay outside its purview. None of these aspects of the *omni* nomenclature can be assigned a place in the visible universe and what cannot be fitted into a place or assigned a velocity and has no dimensions can't be systematically parsed.

Chaos too defies method and even description except for generalities like *confusion, disorder, unruliness.* It comes at all levels from scurrying microbes to scattering galaxy clusters. As Ian Stewart remarks (2016), "In mathematics, 'chaotic' is not just a fashionable word for 'erratic and unpredictable'". It can, however refer "to *deterministic* chaos, which is apparently irregular behaviour resulting from entirely regular laws" (119). Scattered debris is made up of highly regular microcosm units placed where they are by gravity and momentum. They became debris by equally regular laws of physics and hence are both determined and chaotic. The sheer number of objects from small to large moving relative to one another prevents their being mapped or translated into calculations except in local units and in overall estimates where algorithms and methods can be used to make estimates but not to take exact measurements. The spiral shapes of some galaxies and the filaments of galaxy clusters contain large voids and webwork lines of denser configurations that thread through mazes like thin capillaries, again under deterministic laws. Not all regularity in other words is perfectly regular. Gravity pulling from thousands of masses puts wobbles in orbiting satellites and hurls asteroids and planetesimals at one another.

Another reason for the limited fraction of high probability theory under command is that what one discipline finds doesn't automatically carry across to others. That is a disciplinary fracturing in response to nature's multiplicity. It is up to natural philosophy to put the pieces together as best it can. The rules by which it does so aren't as methodical as algorithms, merely the protocols of argument, and argument has semantics to deal with in addition to information and theory. Assuming that close scrutiny and tests produce the most reliable discourse we have, natural philosophy can at least use them to set the limits of what is possible, and that is where it has a legitimate dispute with simplistic stories. It can't credit rituals with real power where the results don't show except for psychological and social ones. Ceremonial occasions and ritual oaths have psychological and legal authority, which is their purpose, but nothing stirs abroad as a result of a sacrament. Whether a spirit is summoned by seance or ritual, from the naturalist perspective the summoning is specious.

The dubious nature of ritual efficacy shows in several ways, the contradiction of one by others, the intangibility of the addressed powers, the lack of a demonstrable connections between what happens and the appeal, and contradictions between what natural history manifests and the characteristics proposed for the operative powers. Whatever Aeschylus and the oracles say, if what happened to Xerxes' fleet off the coast of Athens was due to weaponry, the courage of the Greek navy, and the threat deep water presents to sailors who can't swim, it wasn't Zeus who destroyed the Persian fleet. He may have been summoned but he didn't answer, precisely what Hotspur says in response to Glendower's claim that he can "call spirits from the vasty deep": "Why, so can I, or so can any man;/ But will they come when you do call for them?" (*I Henry IV*, 3.53-56). The question as to whether they do or don't has been batted around for ages without arriving at a positive answer with anything like convincing probability. Prayer as one form of ritual has the same difficulty in the devout knowing whether or not it raises a response.

It will do no good for Glendower to argue his case. His conviction of contact with spirits has no evidence. That is something different from entertaining logically contradictory concepts, which can be handled on the field of disputation and doesn't depend on spirits. Logical contradictions are related to other violated or ineffective procedures. When one thing doesn't follow from another the supposition is supposed to be discarded. Violations of that principle fall more in everyday life and in philosophy than in science or religion. For instance, for a time in Greek philosophy under the influence of Parmenides and Zeno, it was impossible to follow the proposed argument and stay alive at the same time. Given the paradoxes Zeno made famous, philosophers had to keep them apart from experience. They had no choice but to assume that motion took place, yet in line with Parmenides Zeno pointed out that an infinitely divisible space couldn't be crossed whether the measured distance was an inch or a mile. Yet like everyone else they continued to put one foot in front of the other. Their hearts did beat, and the birds did fly. Not only was motion possible but as Heraclitus pointed out nothing ever sat still. Making space, mass, and time *not* infinitely divisible but composed of indivisible units went some way toward solving the problem, but not all Greek philosophers were atomists. The ones still commonly read mostly weren't, yet they couldn't avoid the Zeno paradoxes unless they went to a neoplatonic extreme and declared sensory experience to have only a shadow existence. Proposing minimal indivisible units doesn't completely solve the problem even for atomists, since motion has to cross a space between units, and it is harder to imagine an indivisible unit of space or time than one of matter. Binding space to time and assigning such units to both as spacetime made a plausible solution. Empiricist or inductive measurements of motion and of mass use infinity only to set a boundary for finite units. In that regard an infinite Everything is the same as an infinite Nothing, the *thing* being what is missing. Hence it remains absolutely true that absolutes have no place. The absolute powers of the *omni* nomenclature can make no contact with measurable quantities without being directly responsible for what is fractured, scattered, and painful. That is out of character for any of the proposed supreme powers.

Contradictions

Philosophers not committed to empiricism and natural philosophy are caught in a bind. Consider Montesquieu going in a couple of pages in *The Spirit of Laws* (1750, 1977) from nature and mankind under divine law to war as the first stage of human aggregates. As soon as people enter "society, they lose the sense of their weakness, the equality ceases, and then commences the state of war" (103). Without blinking he derives "unavoidable laws" from God and the state of war from the first law. *Therefore God mandates war*–surely not what he intended to say, although if he did it would be in keeping with the many war god mandates issued to the Hebrews and others in the chronicles of the ancients. The war is an empirical fact. The divine law is a belief derived from revelation. One way out of the dilemma might be to claim that mankind in a state of society is different from mankind under divine law, which with original sin ceased being effective. That would make for a plausible escape if it weren't for frequent biblical commands to make war and the conflict between natural history and a creation that featured the Garden of Eden. Hobbes along with many others of his time encounters a similar problem concerning providence. If it regulates tides it must also be in charge of tsunamis. Another way out of the dilemma is to put limits on the overseeing power, which could create only the best of all *possible* worlds, the solution chosen by Leibniz. But that qualification of *omni* terminology was too serious for most theologians and philosophers to accept and is inconsistent with the history of revelations claimed by prophets, apostles, theologians, and various ecclesiastical hierarchies. Any creator that could make galaxies in the hundred billions should be able to make a satellite orbiting a star less excruciating than the one we know.

Equally in trouble is the more recent notion of Nicholas Hagger (2008) in *The Rise and Fall of Civilizations*, that everything civilizations have accomplished is owed to Light. That transfers knowledge of it to revelation and removes the need for methodical observation. When crops grow, Light grows them, though it apparently doesn't send locusts to devour them or blight to shrivel them. In tracing the transmission of Light and its history around the world with ingenuity and considerable learning, Hagger attributes some of its operation to conquests. Hence much like Montesquieu's sequence, he equates divine Light with war and the imperial ambition of chosen people. The ancient pantheons managed to transfer some of that barbarity to militant gods such as Anat, Ares, and Mars, though Zeus and Juno weren't thereby free to be kind and wise. To be aligned with the nature of things, Light either needs the co-rule of Darkness or must include darkness within itself, which would be as riddling as entertaining motion and the impossibility of motion at the same time. The problem arises from not letting human nature build the civilizations and nature grow the crops and shrivel them by the usual procedures.

The compartmentalizing needed to entertain conflicting systems allows what David Hackett Fischer (1970) calls false analogies and the fallacies of composition, the latter reasoning from a member of a group to the group itself (219), a variety of substituting parts for wholes, one of the more common violations of due procedure. A partitioned mind conveniently forgets counter evidence, as for instance in Adam Ferguson's militaristic view of civilized virtues (1767, 1995). In praising the courage, honor, and affection warriors feel for one another, he neglects the same soldiers' obligation to kill enemies, not to mention bickering within ranks, death by friendly fire, and collateral damage in which innocent bystanders too are killed. That might seem to be less of a problem before missiles and saturation bombing, but war of any kind frequently unleashes barbarity against non combatant populations. As part of that avoidance of the main function of troops, no bodies litter Ferguson's battlefields, no widows grieve, and no children are left orphans, a persistent omission in military recruitment campaigns that emphasize teamwork, training, and dress uniforms. The causes under which armies operate can be defensive and possibly justifiable or aggressive and probably unjustified. How good or bad the individual is depends entirely on the cause for which he fights and how he handles victory. Its abuse is one of the hazards of warfare. The fighting itself, like the sword or the canon, is an instrument of intent, and intent comes good, bad, and indifferent. Japanese troops in China weren't heroes to the Chinese, and German troops not courageous and honorable to Russians. American troops weren't heroic to the North Vietnamese or to the collateral damage victims of Iraq.

In failing to trace the flight of the canon ball to its destination and to probe the mind behind it, Ferguson renders distinctions between self defense and aggression null and void, commonly the case also in heroic literature and in piracy as encouraged and rewarded by a nation's admiralty. Avoidance of realism characterized most battle literature up to War I, at which point field reports, trench poetry, and eventually documentary camera work reduced the filtering of representations. Viewers of the movie *All Quiet on the Western Front* (1930) didn't have to visit a battlefield to find out what one looked like in trench warfare. At the top of the chain of command the chaos wasn't deterministic but voluntary. Extreme remoteness or indifference was necessary to avoid seeing the results of atom bombs in Hiroshima and Nagasaki, in that case arguably a repercussion. A much larger fraction of the truth is available in the modern era of books, newspapers, magazines, movies, television, and the internet than was available in preceding eras lacking in both information and the capacity to send and receive it.

In the General Vicinity of Truth

We deputize words and numbers to stand in for things and quantities, but that doesn't cure the systemic mismatches glanced at earlier. Two common nomenclature difficulties besides interacting components in confusion are that, thanks to incessant change, formerly intact objects leave behind their assigned names, and where they aren't simple the names cite only a feature or two of the total. That isn't subject to procedural correction. Any category title has to select what is important without becoming misleading as Ferguson's ethics of war does in choosing a byproduct and ignoring what is primary. In commenting on the proliferation of biological forms in *The Diversity of Life* (1999), Edward O. Wilson (134) calculates how much shelving it would take to house books that devoted just a page each to known species. Individual variations beyond enumeration would not be included even then. *Johanson* comes to mean much more than *son of Johan* over a lifetime but nowhere near all that Johanson contains or has done. Representing merely a single fleeting moment is as daunting as snagging a particular bat out of the horde exiting time's cave. Degrees of likeness among designated kinds vary. A *community* is a small group of intimately related lifeforms. Generalizations about them hold reasonably well. A *guild* is a set of like species, obviously less uniform in its membership.

Nature's heterogeneity spreads randomness across an area but not to the degree of the pre-creation chaos that greets Milton's Satan, about to set forth to spread confusion of his own. At that extreme no point of view or means of investigation yields truth because none exists where scattered units have no relation. Chaos has for its governance the anti-god figure Chance whose 'rule' is a parody of divine command:

> Before thir eyes in sudden view appear
> The secrets of the hoary deep, a dark
> Ilimitable Ocean, without bound

> Without dimension, where length, breadth, and highth...
> Are lost...
> *Chaos* Umpire sits,
> And by decision more imbroils the fray
> By which he Reigns: next him high Arbiter
> *Chance* governs all. (*Paradise Lost* 2, 890-910)

In the ontology and metaphysics Milton is assuming, chaos and chance are alternatives to obedience to assigned place. Breaking just one rule challenges the designated order. To disobey the source as Adam and Eve do is to undo the structure, which in a marital couple also reverses husband/wife hierarchy. Chaos is what the theological tradition proposed everywhere except Heaven until the Word speaks, the stars gather, and earth comes forth with its creatures. The conflict with natural history doesn't need elaboration. To fit the myth, the actual species evident in natural history have to be limited and radically changed in character. Satan succeeds in returning just a segment of the creation to confusion, though ironically he needs divine help to do even that much. A retaliatory deity is a necessity to make the fable match history. Something similar in attributing a hard life to humans is true of the first book of Ovid's *Metamorphoses* where mankind reduces the golden age to an iron one.

Timing and point of view are critical to the answerability of names to referents and to what we seize upon as at least momentarily true perception. Take a blossom and the word *beautiful* applied to it from one perspective. From far off its best appearance shrinks to a speck and disappears. Under a microscope, it disappears in the prominence of parts. To be perceived even from the one advantageous angle and distance, the bloom must be frozen in time. Its beauty might not be there at all to someone induced by custom to see blossoms as some see snakes. Where the pace of change is slow and the perspective stable, the nomenclature seems more appropriate than it proves to be in the long run. The main exceptions to partial and fleeting citations are abstract categories and idealizations. We set them up as templates as if they were good for all times, whereas in reality they are only comparatively more durable. It is so difficult to imagine perspectives other than those of the moment within our normal sensory range that it takes imaginative exercises like George Gamow's (1965, 1993) to jar us loose.

The senses and the brain as a combined mechanism put still more obstacles in the way of accounting reliably for fractional reality. Consider what happens when we encounter a familiar object. Light enters the retina, strikes the optic nerves, and is decoded in a brain stocked with impressions from similar things. In Howard Bloom's (2000) energetic narrative, cells reject most of what comes in and "diddle mercilessly with what's left, transmogrifying the photons of which light is made into pulses of electrons and bursts of unpronounceable chemicals like prelumirhodopsin. They fiddle with the contrast, tamper with the sense of space, and report not the location of what we're watching, but where the retinal cells calculate it soon *will* be... The eye crushes the information it's already fuddled, compacting the landslide of data" (66). So it goes with the nervous system, the brain, and memory. Before we have finished with one impression, another has already taken its place. Both sink into memory's faulty storage where alterations and blending take place out of sight.

Whatever is out there we can still call reality to keep the distinction between fact and fable, but much of our standard nomenclature doesn't refer to specific parts of it. Take something as ordinary-seeming as the moon riding down in the west. Its name is a rough approximation used for a variety of satellites orbiting other planets. What exactly is this thing? Consciousness is full of stored impressions of romantic moons and the moons of half-remembered poems and songs. Romans once appointed the huntress Diana as its goddess, but she long ago went into hiding from all but scholars of ancient myths. Dismissing these moons and moving on helps us get at the physical object but only in its orbit and at the visible surface from a given angle. If we were to transport machinery to its surface, drilling and sonar probing would get deeper but produce only samples. In likening the moon to similar satellites we get further but still nowhere near what the integral object is. The moon is real. The moon is an imposter.

Beginning, Middle, and End

That definitive endings to any very comprehensive narrative aren't realistic has never prevented authors from tagging them onto world history, sometimes in vividly imaginative ways. That an apocalypse is *poesis* in general category is clear enough, since the ending it proposes has a judgment that applies justice to everyone after applying it indiscriminately across the animal kingdom. Such an ending has narrative advantages. It gives history a direction that includes a personal summing up, and it shrinks an immense universe to fit mankind, which in effect imposes a literary protocol on history and moral philosophy on natural philosophy. Including a personal judgment has the effect of humanizing in retrospect everything that leads up to it. Briefer segments of a realistic narrative can be given more credible beginnings and ends up filtering out anything that doesn't fit. Events coming in from all angles appear only at their intersection with the central story. A life story begins with birth and ends with death without having to account for intersecting biographies. That is the usual way of intersections in artful narratives that don't take detours. It is decidedly not the way of natural history and its master narrative where some intersections alter courses that then go in the new direction for billions of years, as in the encounter of earth's mass with the sun's and its resulting satellite status.

Like the selected parts of a biography, other narratives too have natural contours even if their authors situate them in something more extended. Take Verlyn Klinkenborg's *The Rural Life*, reporting on a particular moment of a particular year: "A light afternoon breeze now carries the tentative bleating of crickets and the hush of leaves in the trees, sounds that seem to advance the season a month or two. Recently the sun shone for five straight days—a feat virtually unexampled in the past calendar year—and high temperatures reached the seventies. By the third day of steady sun, a cautious delirium had spread among the damp-stained residents of the valley towns nearby The root of the New England character is incredulity, a state of chronic-weather-induced heartbreak, and this has been the kind of slow, cold spring in which that character was formed" (69-70). Behind the exaggeration is an awareness that weather isn't always fascinating to the planet's most jaded species and has a way of seeming more capricious this year than in any other on record. The intersection of climate elements has directed the weather hither and yon and created seasonal patterns. The coming of spring marks a turn in *The Rural Life*'s month-by-month progression and says in effect, 'what the author has been heralding these many pages has now arrived, and it is an anticlimax'.

That blighted spring would lie adjacent to what Northrop Frye (1957) calls the mythos of winter except that *The Rural Life* is predominantly more reverential than ironic. The train of years leading into this one and away isn't needed for the immediate purpose except as an implicit standard for how seasons at this latitude should go. The structural point is that despite the regression and progress, a calendar year makes a natural unit. Segments of PBS' *Nature* series often come in that format. So of course do financial matters except subdivided into quarters. Days and years are natural units. Weeks and months are clock time and calendric grids. Decades, centuries, and millennia are multiplications of conventional units. Clock time in minutes and seconds is enclosed in the natural daily cycle. The grid of longitudes and latitudes is laid upon the planetary sphere imposing geometric regularities that in representations overrule canyons and mountains.

Setting goals and providing endings are similar contrivances with some justification in natural regularities. With a few lapses Darwin's version of evolution concedes that degree of randomness, but not all versions of Darwinism have done so. They have instead insisted on directional intention without defining a conclusion to it. In scrutinizing crystal, for instance, John Ruskin thinks (in *The Ethics of Dust*, 1875, 2006) that the evolutionary narrative is dedicated to progress, hence the paradox, *moral dirt*, another of the conflations like deterministic chaos and regular irregularity. Crystal's geometric triangles, hexagons, and octagons have emerged from less orderly matter. Dust makes a "continual effort to raise itself into a higher state." Sand is on its way to "smooth knots of sphered symmetry," to opal. As slime dries, it withdraws from anarchy and infuses "a finer ichor into the opening veins" (163). Even society could improve if it would abandon cutthroat competition. As with crystal, so with the social evolution from barbaric life to civilization, except that in Rousseau and again in Marx and Engels civilization got derailed and needs a revolution to set it back on course. Needless to say, Ruskin's evidence selection has to be quite strict. Evolution produces many imperfect things not on the way to anything better. Sand processes into much sandstone and little opal. Most dust hasn't gone anywhere except where the winds direct. Depending on what comes in contact with them, dust and gas in space may take shape in rings around Saturn, collect into a rock planet, or remain chaotic indefinitely.

Stuart Kauffman's more plausible biological variant of progress starts with the earliest bacteria and converts evolving diversity into a program designed to make humans feel "at home in the universe," or as the title of another of his books puts it, on the way to reinventing the sacred. That too depends on rigorous evidence exclusion. Humans can't really consider themselves at home in a universe that is mostly uninhabitable and from their location stretches in miles some 14 billion multiplied by 5,878,499,817 in opposite directions. One arena is incompatible with another, as biology is with cosmology except as its derivative. Since the total entropy of the projected end means the cancellation of information, the interim becomes not a programmed progress for any theory of advancement, only a very prolonged state of limbo.

John Burroughs' notion of accumulated form and disintegration in "The Grist of the Gods" (2001) comes closer to how natural cycles of buildup-breakdown buildup actually work:

> The soil which in one form we spurn with our feet, and in another take into our mouths and into our blood–what a composite product it is! It is the grist out of which our bread of life is made, the grist which the mills of the gods, the slow patient gods of Erosion, have been so long grinding–grinding probably more millions of years than we have any idea of. The original stuff, the pulverized granite, was probably not very nourishing, but the fruitful hand of time has made it so. It is the kind of grist that improves with the keeping, and the more the meal-worms have worked in it, the better the bread. Indeed until it has been eaten and digested by our faithful servitors the vegetables, it does not make the loaf that is our staff of life. The more death has gone into it, the more life comes out of it; the more it is a cemetery, the more it becomes a nursery; the more the rocks perish, the more the fields flourish (2001, 135-136).

Some of those progressions a pessimist could just as logically reverse. The more productive the nursery the greater the cemetery. The complete formula gives us both growth and decay among lifeforms and buildup/breakdown cycles in material composites. The transformation of minerals into grain and of grain into a loaf is followed by further transformations in cycles of composition and decomposition. Soil passes through many generations of worm work, assisted by insect work on vegetation. The grist can also be compressed into sedimentary rock or put through the metamorphic process of gneiss. In any case the present form of living bodies and of mineral matter is "only one of the infinite number of forms that matter must have assumed in past aeons," Burroughs concludes (140).

That isn't the entire truth either, but it is a sizeable fraction and not misleading as myths with decisive beginnings and endings are.

One way to look at both yearly and life cycles is as regularized transfers of matter and energy, like dragging and dropping an electronic file from one folder to another, not really an algorithm but a patterned process subject to enough variables to be messy at times. After leafage has absorbed solar energy, the nibbling cottontail converts it into flesh and bone, one chemical marvel following another. What has made the rabbit hop then makes the coyote run. Ovid's *Metamorphoses* come reasonably close to that transformation. We don't take umbrage at the first of these chemical transfers any more than we do at earth's collecting and transforming solar energy, so why take umbrage at protein, calorie, and mineral exchange? The fly and worm get their turn with the coyote before leaf, root, sunlight undertake another cycle. "The more death has gone into it [and into disintegration] the more life comes out of it." Except of course with the crumbling and reassembly where nothing living is part of the cycle, which is everywhere so far discovered except around a thin, broken curst of one planet.

Essences

Sooner or later nearly every branch of philosophy encounters the question as to how much our verbal and number systems mirror nature and how much is an organizing convenience. That behind objects are permanent forms as opposed to near replication has been one of the truly stubborn illusions. It is based on a simple mistake: taking semantic abstractions and math calculations for templates, perhaps with an intelligence like the Christian Word or Plato's Demiurge dispensing them. The near replication of mechanical and biological combinations isn't the problem but the question as to which brings it about, natural laws or creation. Plato's essences and Aristotle's forms, Augustine, Thomas Aquinas, and others joined to the Word for an animated version of essence theory. Plato's concept of resisting, form-spoiling matter requires reducing the power of the application, again not usually acceptable to theologians who apply *omni* nomenclature to the source. It also doesn't work well with living, growing, and decaying life that doesn't have an essence or perfected bodily form. The history of matter does of course follow patterns, but every moment of change as well as every species would need its own form or a perfect version to make essences work. Quite a few poets as well as philosophers have assigned high value things endurable form, especially where love is the subject. In common terms aside from philosophers and poets, people talk of eternal souls as if they, too, had a permanent form independent of experience, a core self present as much in the infant as in the adult and capable of existing unhampered by the lack of sight, hearing, and touch when the soul leaves the body. At least for the hyperbolic purposes of a given poem love is to last forever. Thus Shakespeare's speaker admonishes himself in sonnet 116, "Let me not to the marriage of true minds/ Admit impediments." Why? Because "Love is not love/ Which alters when it alteration finds," which is to say that love must have durability despite being in the care of people who change. If a presumed lover does change, then he or she hasn't really been a lover. *Love* in Elizabethan terms included friendship and was frequently used in politics to indicate unwavering loyalty to an ally. In reality the actual beloved had parents, grew from infancy, and is subject to the usual cycles. On several occasions in the *Songs and Sonets*, among them "Loves Growth," "Loves Infiniteness," and "A Lecture upon the Shadow," Donne puts stages of love and fickleness in the conventional Platonist context. In "Loves Growth," love is allowed to grow but not decline. In "A Lecture upon the Shadow," love can be "a growing, or full constant light," but its "first minute after noone, is night." In real life, affections can change without plunging completely into darkness, but should Donne's true love falter for so much as an instant (say in a mistress' roving eye), it vanishes altogether. That would have to be the case if it were an essence.

Either the word *Love* designates an eternal verity or no word designates anything durable. "For love, all love of other sights controls," as "The Good-morrow" puts it. Shakespeare, too, finds language undermined if the lasting value of *love* is: "If this be error and upon me proved,/ I never writ, nor no man ever loved." That all-or-nothing postulate underlies comprehensive fictions whatever the key words, *soul, god, truth, beauty*. For Keats in "Grecian Urn" it is the latter two that can't be compromised and must be one and the same. The only way to sanction such an idea is to assign it a mental compartment of its own and not let it make contact with objects, sensations, or intellect. In a poet that is love talk. In a Platonist it becomes philosophy. Progressing from instances to universals is the aim not of lovers generally but in Plato specifically of wisdom (*philo*, love + *Sophia*, wisdom). That is the primary reason for philosophy's superiority to sophistry. It deals with essences, sophistry only with words and appearances. Something, somewhere, has to be stable or a philosophy based on essences collapses. *Its* first minute after noon is night.

The empiricist view is that abstractions like *Love*, the *Good*, and the *True* needn't have corresponding entities any more than paper currency must have gold reserves behind it. In a universe entirely in motion, instability comes in variable speeds as truth in realistic use comes in gradations and applies to correspondences between statements and things. The Anglo-Saxon word *truth*, originally meaning loyalty and faith, eventually became synonymous with *versus* in Latin and *verity* in English, meaning in agreement with or corresponding to fact. Statements can correspond more or less to fact, hence fractional truths, as in saying all men are created equal we allow exceptions that make *all* idealistic rather than factual. Equal in what precise ways is left unstated. The term *stone* is an approximation that generalizes variants without suggesting that an essential stone exists somewhere. Polished by water in a streambed, a given rock speaks of ages of fluvial wear, in no interval of which has it been immune to shedding molecules. That some sand originally deposited at the sea's edge is currently mid continental sandstone tells of even more extended spans and transformations. The continents themselves not only travel laterally but shed volcanic ash, wind-rafted dust, and detritus carried out to sea by rivers and glaciers. *North America* means something different every day. The same with everything in natural and human history.

Pockets of Difference

Until considerable work had been done on fossils, species were assumed to have originated *as*, if not necessarily *where*, they are. That gave them something like an essence carried through stages of maturation. Scholars of flora and fauna knew that species differed by place, but until place was found to be what altered them, the variety lacked a satisfactory explanation. It was left to the whims of creators from Coyote to Tiamat to explain aardvarks and platypuses.

Among the post Darwinian American naturalists who found evidence of place-shaped variants not many were as observant as John Burroughs. When he died just short of 84 in 1921, the essays he left behind were indebted to the romantic poets, especially Wordsworth, but were also aware of evolution and geology. *Curiosities* his observations could often be called, with *love* (not the eternal variety) stationed over some of them: "You must have the bird in your heart before you find it in the bush." The naturalist needs that as well as a "highly sensitized mind" familiar with kinds. He feels challenged by the small, the fast, the well disguised, and the incidental. "The stage [nature] is very large and more or less veiled and obstructed" (6-7). Rambling on foot brings the best observations. Rambling through miscellany makes a better metaphor for discovery than a purposeful search. Curiosity wanders among possibilities that might easily be missed, with differentiated pockets within a region missed almost as easily as the very small and the very fast. Every sizeable area has microclimates that illustrate that principle of relative placement and movement. Pocket phenomena are typical of the variability that evolutionists took some time to understand. Either odd-looking species in Australia were created to fit the specific habitat or in their isolation they somehow changed from parent forms. In either alternative, *juxtaposed unlikeness* and *distant likeness* put kinks in evolutionary models. The political and ethnic equivalents are alliances in pockets distant from one another. Allies in kind may be separated by entire seas and continents. Similarly with pocket ecologies made up of combined symbiotic and hostile species where relative velocity means a rabbit fleeing a coyote.

That disjunctive principle–interrupted proximity and distant likeness–has implications for other aggregates besides dispersed social alliances and species mixed in ecologies. Naming a kind of locality puts language further up on the approximation ladder, as *rain forest* is unspecific in content and cites a general context for many species. In physics luminous and non luminous particles are unlike but presumed to be intermixed. Farmers who discourage the cohabiting of unlike things tame a multi species wilderness into exactly placed rows of corn. "Weeding out" is a prototypical human activity, one of the many ways to alter an environment and make its remaining parts congruent, a cleanup as well as a growing operation. As Leopold observes, wildlife and what are labeled weeds can be given allotted places in the midst of cultivated land as a concession to the wild. The problem isn't hawks and owls or peppermint and flowering spurge but their placement. Peppermint choking out alfalfa isn't what a farmer wants, but alongside a stream it presents no problem. The hawk that kills chickens is bad. The one that gets gophers is good. A proper sense of information- yielding structure puts one inside another like kachina dolls, as inside a living body are organized organs and inside them are clustered cells. In a supercluster of galaxies among an irregular field of them are smaller clusters joined by filaments, inside which are stars and satellites, and so on down through accrued components to subatomic particles. Each layer is structured and hence can become a focal object of an anatomy with characteristic forces of cohesion, movement, duration, and exchange.

People do something similar in joining with and parting from allies. Consider ways of life, modes of production, and customs. Different rates of advance in stone age culture and agriculture have trouble coexisting. As Colin Tudge in *The Time Before History* (1996) remarks, "a farming population is liable to be more powerful, in a crude military sense, than a nonfarming group," a distinction that marks the transition from nomadic clans and tribes to settlements and so on to city states. Agricultural production produces similarly oriented populations however distant their localities: "If your neighbors are farming, you had better start farming as well; otherwise you will be swamped." In brief, "farmers obliterate non-farmers in exactly the same way that human beings… override existing faunas when they arrive on fresh continents" (Tudge, 273). The overall favorable habitat of a temperate zone becomes hazardous for contrasting modes of production and adjacent cultural pockets. The reasons are mainly pragmatic. City dwellers need only walls and a roof overhead, not large acreage, but out in the country herds of buffalo are incompatible with wheat fields. Fences are barriers to hunters and their prey. It isn't in the situation or the heritage of either to work out a mutual survival plan or to reorganize and merge.

Not All Similarities Are Alike

As the science educator Jacob Bronowski testifies, scientists don't usually add up instances to reach precepts. Even Bacon, sometimes thought to over emphasize induction, never held that scientists should rely exclusively on it. As Richard Dawkins (2008) also remarks, new instances come already "enmeshed in a huge and intertwined network of theories" (178). Scientists know ahead of time how an electron or the eye of the fruit fly works. They set up tests not to discover but to verify and fine tune. That isn't an ambling exploratory side of learning but a directed course. Predictable consistency is proof satisfactory to reason but not definitive. An exception might exist somewhere. That theories are often too broadly applied for the evidence is partly what Bacon had in mind in championing induction. Similarities in phenomena lead too quickly to generalization and to theories without verification. Analogies among different levels of placement are especially difficult to apply and need to be monitored. Darwin's insight into natural selection at the species level doesn't apply to societies, at least not in the way social Darwinists supposed. Darwin's own speculations as to how individual natural selection becomes collective are tentative.

One thing those who misapplied biological evolution ignored among other things was the dumb luck factor that deals a poker player a royal flush. The survival of the fittest is true enough, but fitness is quixotic and changeable. When volcanoes alter the atmosphere, surface life suffers while fish survive. Up to that moment, the fish weren't fitter. When the seas warmed in the Permian-Triassic die-off 250 million years go, something–a poisonous plume has been suggested–annihilated 96% of marine life and 70% of land vertebrates. It took 50 million years for comparable land life to return and twice that for marine life. Fitness in the recovery of some species and generation of others was periodically redefined over that extended period. Next time it may be the elephant that survives and the shark that is unlucky. Each member of each endangered species is made up of parts, most of which could be functioning quite well while something in the environment is inimical to a different organ. The toxic plume need only attack that organ to eradicate the species.

Calculating the odds of a species' well-being encounters too many variables to be an exact science. One false analogy prominent earlier than it is now connects nature to morality. That notion survives in concepts of providence and divine uses of nature as an instrument of blessings and curses. Eusebius (2006), to take an early Christian example, believed that the misfortunes of the Jews after the crucifixion, presaged by "wonderful prodigies," was due to

God's reaction to the unrepentant nature of the Jewish character. *Stubborn* and *stiff-necked* were for code words for that. That forty years separated the crime from the punishment wasn't the product of disconnection in Eusebius' logic but of "the benignity of;... all-gracious Providence" (77). Unfortunately for a normal sense of justice, by the time providence got around to reacting the real perpetrators were gone and innocent victims bore the punishment, like a child of the 21st century coming down with whooping cough for an offence Adam and Eve committed. That a supreme being is watching one's every move and keeping score is still a surprisingly contagious idea in moral philosophy disconnected from the evolution of humankind. When exactly the ledgers started being kept in the gradual transition from Australopithecus to *Homo sapiens* isn't usually part of the discussion, since natural history has to be denied to allow the concept.

In other areas as well lingering superstition supplies moral reasons for what comes about by other means. A great many variables go into the lightning that strikes a plane in flight, and no harm was intended by any of them. The overall natural history narrative does connect everything in a causal sequence but with less uniformity as the miscellany grows, as the evolutionary tree represents lineage with close likeness immediately after a branching and increasing separation thereafter. In a *New York Times* article April 8, 2016 by Carl Zimmer, a team of scientists in *Nature Microbiology* alters the tree to what resembles a shredded fan. Another kink in the transfers from one generation to another is that some deviations from an ancestry change faster than others. Of two species with a mutual ancestor one can still resemble the source and the other be radically different. The relation of adjacent and distant parts has to be worked out by tracing DNA transfers in kinds close enough to reproduce. Stage-by-stage evolution can hold fast with a given kind or accelerate its change.

Our means of representation relate variously to the inductive procedure of empiricism and to the inventive faculty evident in *poesis*. A necessary sequence has only one means of progression. Unlike a *specimen*–an actual instance among a number that can be estimated—a *symbol* has a range of meanings and can have several plausible applications. *Parables, fables*, and *myths* belong under symbolic fiction with likenesses to reality without being literal. *Charts, graphs, diagrams, templates,* and *models,* to fill out the representational spectrum, can represent relations among fictional, mythic, and real things. Nothing keeps any of these from being reliable except misreading. The story of Pandora's box has many applications and finds many resemblances. The only positive misreading would be to presume that some such creature actually did open a box and let complications fly out, Hesiod's way of explaining why life isn't simple. We've no problem seeing that because along with the rest of *Works and Days* it is a legend not meant to be taken literally. The Hebrew anthology, New Testament, and the Koran are otherwise in their respective regions of belief. How many such myths of origin *have* actually been considered historical?

Greek entries in the *Oxford Companion to World Mythology* (2005) number nearly 300 figures, many of them responsible for originating something. In their whimsical trickery, the fifty plus myths in Africa (a fraction of the total) suggest more entertainment than actual belief. That is likely true of stories taken seriously by only a segment of the population. That no empiricist tradition existed until disputation went to school in Greek academies and emerged with protocols for argument doesn't imply lack of earlier skepticism.

Why People Took
To Collective Illusions

Chapter Four
Skull Duggery from Lucy to Mythology

Spell Binding Stories

Two reasons for going further back than usual in making the case to reclassify myths commonly mistaken for history: we can't fully understand human vulnerability to ego centered myths without knowing what led to it, and the length of species history is itself a commanding reason not to credit myths of origin. How many still do? A growing number believe it or not according to studies of fundamentalism and polls on the relation of religious belief to evolution. First a basic fact everyone supposedly learns in school: the place of people among the species came in an evolutionary process that had been over three billion years in the unfolding from single cell organisms to four and eventually two footed land species. As to the invention of myths: they came late, when civilizations were forming, though nothing was inscribed or otherwise written for several thousand years after the founding of the first urban settlements. Images and hard object artifacts supply what evidence there is, including pictographs of hunters and prey 30,000 years ago, several millennia after the probable development of proto language into grammatical language. Hieroglyphics and cuneiform clay tablet inscriptions were still a long way off at that point. Alphabets arrived some 26,000 years after one of the suggested times for grammatical language development, 50-40,000 years ago. Cuneiform inscriptions using several hundred ciphers were in use in Uruk by 3000 BCE. The first elaborated myths weren't committed to script until Sumerian culture about the year 2000.

Before we consider (next chapter) the contributions of mythopoeic thought to more systematic deliberation I want to see if the illusions, trickery, and treachery that accompanied language development can't be separated into pre and post grammatical language stages. Presumably, despite a surprisingly varied vocalization in primates, only after grammar and semantics could a hypothetical way of thinking have reached any very elaborate vocalization. Proposing something that might be requires a level of imagination beyond ordinary anticipation. As a category elaborated in language the *as if* at whatever stage it came would have been a tremendous intellectual tool opening the way for a contrary-to-fact doubling of consciousness–what exists and what might.

If we want to go further back than anthropology into zoology to test the range of meanings possible prior to language we can hear warning calls for ourselves from all around the animal kingdom. All creatures want timely perception on their side. They can alert fellow clan and flock members and warn off intruders. Issuing alarms and warnings is inborn and doesn't require language. That is the case also with deception, camouflage, hiding out, and pack behavior based on alarm. These are mainly survival reflexes for which zoologists find both land and sea creature ancestry. Alarm spreads easily and quickly. It would likely have been among the first group gathering mechanisms in flocks and herd animals. Cooperation in other situations parallels group alarm. Going with appearances usually pays off even if occasionally a footing that looks sound proves to be quicksand. At whatever level of sophistication, the organizing function of synchronized beliefs depends on what either looks authentic or holds enough promise to encourage belief. Anticipating the notion of culture-evoked change, Adam Ferguson (1767, 1995) was among the first in print to expand on that with attention to conniving and contriving, in his case ignoring the animal roots and turning directly to the most advanced of inventors: man is "in some measure the artificer of his own fame, as well as his fortune, and is destined, from the first age of his being, to invent and contrive He would be always improving his subject, and he carries this intention where-ever he moves At once obstinate and fickle, he complains of innovations, and is never sated with novelty" (12-13). That invention spreads into all fields including the concoction of what-is-not, the contrary-to-fact. Deception only has to look like candor to be the choice of the moment in the opportunistic seeking of what works or looks as if it will.

Verbal Buildup

Discussions of generative grammar and universal language capacity suggests why we should neither ignore nor exaggerate genetic mandates when it comes to developing communication beyond gestures and calls. The occasion of one such has been a linguistic study by Russell Gray and colleagues that found fewer generative grammar patterns across four family trees of 2000 languages than might be expected from the work of Noam Chomsky (1988, 1991), Steven Pinker (1994), Ray Jackendoff (2002), Joseph H. Greenberg (1966, 2005), Bernard Comrie (1989), and others. Those who have proposed diagrams of the brain point out that when our ancestry encountered certain situations frequently, specialized areas developed to handle them. They became instinctive and unconscious. As Leda Cosmides and John Tooby remark (1994), brains got partitioned so that hearts could pump and livers detoxify. Those domain-specific capacities (where to stop in listing them is one of the problems) aren't necessarily active, but they stand ready for "*evoked culture*" (108). *Evoking* would also do for contagious ideas that elicit similar responses throughout a group.

Language acquisition in children also gives us insight into the infancy of language development. That applies to other parts of communication as well from facial expressions and gestures to semantics. In children awareness of make-believe increases gradually along with the rest of communication. It is a rare child who lacks a sense of humor and capacity to play roles, and a rarer one still who can't be deceived. Since accelerated language learning lasts only so long, its tapering off faster than other kinds of learning implies that it is activated biologically. Before the accelerated capacity fades in youth it adds several thousand distinctions to references: "Language is not a cultural artifact that we learn the way we learn to tell time or how the federal government works," Pinker argues. "Instead it is a distinct piece of the biological makeup of our brains" (18). It is a vital part of sociability and information exchange.

The interactive ethology proposed by Harvey Sarles (1977) contends similarly but with emphasis on mother/child extra-grammatical interaction, usually the most important environmental factor for infants. Infants have a non-linguistic sense of play as evidenced in peek-a-boo games. By age two, a child will ask in mock disgust, "are you joking [yet again]," indicating an awareness of simulation as a playful twist in relations. In playhouse situations the games bring cooperation in make-believe, an early step toward the nomination of a skit writer or spin doctor. The capacity to detect deception develops in tandem with the capacity to deceive. More to the point, it had to have done so in species development all along. Deception exploits grammatical language but has a life apart from it. In animal intelligence it preceded language by a considerable span in making use of body language in covert theft, in stalking, calls, and in camouflage.

Terence Deacon in *The Symbolic Species* (1998) offers a reason not to assume anything very specifically genetic behind childhood's accelerated language learning. Vocabulary has to be learned the hard way. The grammar, syntax, and other features of language mayhave been user friendly to begin with, much as software rewards what users are inclined to do in other contexts. If the main features of grammar fell into place more easily than word-by-word semantics, our ancestry might not have needed much specific interior programming to make use of them. Phonemes and the meanings of specific words are less likely to have hitchhiked on anything else. Because names in alphabet language are arbitrary, attaching specific ones to specific things resembles nothing prior.

Whenever its beginning, vocal communication beyond primitive proto language evolved in response to the need for social exchange and probably to the pleasure of it. Nothing comparable was possible where syntax and semantics were missing. Robbins Burling's account of what grammatical language eventually accomplished takes deceit seriously as the bonding of the recipient's mind to a simulated one (2005). Only by means of language could early humans "*accuse, advise, answer, challenge, claim, demand, deny, discuss, describe, encourage, explain, flirt, insult, invite, joke, learn, lie, negotiate, object, promise, pretend, question, reject, request, refuse, teach, threaten, warn,* and *woo*" (211). Burling's talking apes, Merlin Donald's minds so rare (1991), Jared Diamond's third chimpanzee (1992), and Terrence Deacon's symbolic species (1997) play competitive games that enlarge the insulting, coaxing, objecting, deceiving, refusing, threatening, and warning aspects of language. Add lying and storytelling to the list and misinformation to histories of information. The larger a given collective, the slimmer the chances of an imposter being detected, the less chance of incurring a penalty, and the better the ratio of reward to risk. For that reason and because of the advantages of cooperation, not much trickery was likely in small bands of people. Bickering and competition, yes; cheating and calculated deception probably not much. Wholesale deception wouldn't have paid off until trust could be more safely violated. Probably status would have had to be sufficiently complex to encourage identity manipulation. An imposter comes at the far end of role playing artifice, the middle and opposite end of which are the more usual and acknowledged roles of chieftain, matriarch, drill master, teacher, pupil, lover, friend, and other selections from the common repertoire. A subordinate could pretend to be a chieftain only in play. That doesn't of course prevent contests for prime position.

We encounter subjects, objects, and predicates more often than this or that object. We also take readily to hierarchies of abstractions because we encounter *animals* more frequently than *mice* or *eagles*, and the latter more often than the mouse in the garage or the eagle in the aerie up on the cliff. Because experience recurrently sets objects before us to be acted upon, constructions like "I pluck this berry" come naturally. Sensori-motor processing in primates contained a precursor of that actor-action-object structure as in a bonobo reaching out to seize its food and cultivating a brain-nerve pattern. The reaching out doesn't imply self consciousness, merely an operative distinction between doer, deed, and object.

The S-V-O example is simplistic, but it illustrates a principle of some importance. The monkey that climbs a tree to escape a leopard is not thinking "I the monkey see a leopard and must scurry," but the foundations of such a S-V-O utterance are there. Creatures reaching for bananas evolved into *Homo sapiens* making statements about reaching for bananas. If every animal kind has a latent concept of an agent, act, and object, it isn't a large step to assign these category concepts. Attaching adjectival qualities would be less common. Grass is *appealing* to the wildebeest and the agent *eager* or *sluggish*, but for someone to indicate that requires a more developed vocabulary than pointing at objects and making a sound. The substitution of symbols for things and the construction of longer sequences and narratives would follow only in due course. How we learn language doesn't directly apply to how we come to mistake as real whatever has a name, but acquiring applicable syntax and semantics enables illusions more complex than thinking the crash of a falling branch in the woods is an approaching bear. The authenticity of elaborated truth-saying prepares for crafty simulation. Its motive may be no more than crying wolf just to see what happens, triggering the alarm system that comes built into nearly every creature that swims, crawls, flies, or walks.

Genetic Programming

The implication is that if some aspects of language learning are genetically implanted so might verbal craftiness have been as the playfulness and deceit in animals suggests. How it is transmitted from one generation to another is a subordinate question that doesn't necessarily have to be raised in this context any more than how other automated reflexes are. Some evolutionary biologists suggest, however, that *genetic hitch-hiking* and *genetic assimilation learning* can speed up the process of making learned behavior instinctive, and since language increases the range of storytelling and therefore deceit, it might be possible to say that humans became more instinctively vulnerable to deception after the development of semantics and syntax. Certainly the range of what could be conjured out of thin air increased dramatically and could have seized upon the emotion of the alarm call. That idea is appealing because it suggests an incremental learning capacity in which a gain however small makes the next step easier, and that one the next. In hitchhiking, as Michael Corballis (1991) explains in *The Lopsided Ape,* genes on a DNA string lying adjacent to those with survival value may be selected by proximity (14). An incidental dragged-along change can be put to use whether or not it has great value itself. That is akin to the transfer of a general ability to a specific application, of agency, for instance, to supernatural agency, in that case a transfer of something familiar to something unknown. Again, as a psychological matter a known truth that resembles something mistaken can give the mistake loaned credibility, the reason demons and sky figures have recognizably human emotions and angels fly like birds. Thus even in the *omni* terminology, a deity in charge of billions of galaxies and infinitely beyond anything human seems real when projected as a father figure or a goddess. As Stephen

Jay Gould, too, points out in several places, evolution itself has an accidental element, which he and a colleague liken to spandrels in architecture, the non supportive space available for ornament adjacent to supporting arches. By striking it lucky, genes move from superfluity (spandrel position) to the line of duty (arch position) where their survival is more assured.

Richard Wrangham and Dale Peterson (1996) have still another way to trace how one modification in biological structure boosts another. A drying African climate favored woodland apes who could use forelimbs to dig for waterlogged tubers and carry them to safe places for consumption (54-57). Legs moved, arms dug and carried. The survival value of digging and weight-bearing arms depended on the buried food supply. Having arms free then found a multitude of other uses not related to that one. Whether that notion is valid in this particular case or not, Darwinists generally assume that accidental adaptation and selection do take place. Evolution makes use of one development to support another. Being a land creature is a lucky survival factor when an underwater plume kills sea life, but lungs and legs have functions other than avoiding toxic plumes. In loaning its services to lies and illusions language took on functions not originally included in communicating information.

Genetic assimilation learning, an alternative way to speed up biological change, converts learned behavior into genetic inheritance. That is harder to see and may look too Lamarckian to be reliable, but John Maynard Smith and Eörs Szathmàry (1999) find it possible on statistical grounds. A certain number of neuronal switches must be set correctly for advantageous behavior to pass from one generation to another. Where a favorable mutation depends solely on genetics, having nearly all the switches set right is no better than having only a few right. A mutation will disappear in the next generation (166). However, if learning sets some switches correctly, the combination of those and genetically set ones would more often add up. An advantageous learned trait would more easily work into hard wiring (166). It may or may not be that language propensity and the scheming embedded in it evolved faster that way, but by that or some other means articulate speech did develop in a comparatively limited span after proto language had stalled in *Homo erectus* perhaps for as long as a couple million years. Only a limited range of advancements in mental and skill exchanges through teaching was possible without the syntax-and-semantic breakthrough. With it, in a comparatively brief time *Homo sapiens* could have grown into *Homo ludens deviosus insidiosus*. The enabling mode of production that grouped farmers in settlements brought myths, illusions, doctrines, rituals, and aggravated adversarial polarities along with the arts and sciences, monumental architecture, and *poesis*.

The Clever Strain

Invention, a broader and more inclusive term than *deception* or *illusion*, had, and obviously still has, survival advantages. What was needed to turn its deception branch conspiratorial was mind reading. That leads us back into zoology, where deciphering what others are thinking makes use of signs and sounds other than words. The costs of deception and gullibility in terms of eroded trust apparently weigh less in certain species than the benefits of cleverness overall. Crediting the incredible, to put that negatively, must be less damaging than too little curiosity and invention. To go far back into the Cenozoic and Mesozoic, the evidence of individual animal cunning is substantial, less so for cooperative deception. The Sauropsid neighborhood (Dawkins' 16th rendezvous in *The Ancestor's Tale*, 2004) provided a distant kinship with turtles, "snakes, chameleons, iguanas, Komodo dragons, and tuataras" and "the dinosaur-like reptiles or archosaurs." (254). In that reptilian branch of evolution, both hunter and prey developed tricks to supplement fleeing and fighting. Dodginess combined physical agility and hyper mental activity. Anticipating the next move remains a good trait to have at any level. The bipedal branch of that didn't get underway until after its separation from apes, but the prior separation of apes from monkeys also had something to tell zoology and anthropology about mammalian ancestral trickery. Drawing on zoologists, a 2009 PBS *Nature* presentation entitled "Clever Monkeys" depicted a surprising range of behavior in distant cousins of ours native to Borneo, South East Asia, the New World, Old World, tropical forests, African savannahs, dry forests, and river habitats. Marmosets farm trees for sap, for instance, leaving wounds to bleed and returning for the harvest. That creates durable property attractive to rivals. Theft and conflict are the results, thus monkey antecedents to farmers and brigands, in that case distinguished by tribal cooperation on both sides.

Humans aren't the sole signaling species either, and with signals come the means of deception and its exposure. Ways of banding together multiply. In crossing distances, signals alert numbers. Troops of eight different species of Africa monkeys from Sooty Mangabeys on the ground to Red Colobus and Diana monkeys in the canopy decipher one another's calls, combinations of which number over 120. These can be rearranged and thus require the deciphering of rudimentary syntax. The eight species "behave as if hearing a sentence," the PBS commentary concluded. The cross species calls are numerous and mostly intended, as when monkeys in west India see a lion that the lion's prey doesn't see and issue an alarm call. Off go the potential prey that recognizes the warning. As to more elaborate communication on the way to pedagogy, macaques in Sri Lanka practice leaf culture passed down through learning, thereby anticipating medical school. The practice of animal medicine is prominent enough to spawn research of its own under the name *zoopharmacognosy*, or animal self-medication, taught, not individually discovered. Since neither the knowledge nor the drugs themselves are being sold for profit, no chicanery is involved at that point. Value has to be recognized before scheming begins. Alchemy is as far in the future as are the shaman and the soothsayer. Still, the negative rides along on the positive. Patented medicines and placebos had their day along with 18th and 19th century legitimate pharmacy.

Because false signals always presuppose genuine signals, an underling in a hierarchy wishing to poach privileged food can issue a false alarm that diverts the attention of food guardians. It can then filch morsels reserved for its superiors. Clever monkeys have been caught in the act on camera. Some practice at that level of deception and thievery was in the works a minimum of 15 million years ago. Tanzanian fossils of a line of old world monkeys leading to apes and humans have been dated at 25.2 million years. Characterizing that too falls to zoology in whose province the roots of human behavior lie in chatter, the bonding of clans, and some interior clannish trickery. Theft requires a degree of mind reading and sufficient detachment for the thief to want to trick someone, in other words the makings of empathy in understanding and of individual separation sufficient to overcome it. Thievery carries a guilty conscience because of that betrayal of trust. Among accounts of several kinds of animal deception in "Tools of Deceit" (141-172), Marc Houser (2000) tells of fake alarms detected by Charlie Munn in mixed-species flocks of antshrikes and tanagers in Peru.

Where thievery, trickery, and aggression are scolded, the reprimand strengthens the conscience and intensifies the guilt, which was strong enough in Japanese shame culture that troops in War II sometimes chose suicide over capture. Hauser doesn't think we can ever know exactly what another animal is thinking or feeling (256), but that too easily becomes an all-or-nothing principle. In line with other fractional knowledge, we can know something about how animals feel from how they act. A would-be trickster in the animal kingdom can see that much as well, as any such must if the deception is to pay off. It must be promising to be attempted, and its detection must be successful enough of the time to prevent a serious erosion of trust. That fits the profile of human crime, which can take on animal like creepiness, as in a video recently of a skulking gunman taking cover behind a car and opening fire. The bent over posture and careful stealth, the hooded concealment, and the crouched run from the scene looked animalistic in video pantomime. Prototypes are common in the stalk-and-strike animal world and become familiar to children playing hide-and-seek. Leaping out of concealment with claws made of fingers is a simulated version of something quite real in nature.

A relatively advanced level of sophistication appears in geladas (a form of baboon), the "chattiest of monkeys" on the savannah. The chatter diffuses tension raised by group numbers and the dominance of harem masters. Bluffing shows of teeth and threatening gestures forestall conflict, though group and individual skirmishes do break out. Where teeth and claws are up to the task, even fatalities are possible. That the displays are more ferocious than the intent to harm is a beneficial form of pretense. Seeming innocence and a poker face, too, have been around a good long while, and their function isn't always to get away with something forbidden. Bluffing comes natural to many lifeforms and is sometimes performed by bands of them. In group hunts and defense, flock oriented animals present a united front of hoofs, claws, teeth, and noise. That can be either play acting or tactical. In case an actual fight breaks out the coordination has been rehearsed ahead of time. It takes a propaganda barrage and large scale brainwash to fill avenues and squares with uniformed marching troops eight abreast. The sheer number in the demonstration emboldens the troops and intimidates potential targets. Regimented fascist and communist demonstrations in the 20th century came in numbers exceeding even those marshaled by holy wars.

Sportiveness and Playing Dead

Myths of origin aren't necessarily wrong or misleading about the sources of deceit and its link to competition and hostility, since after all the main purpose of deception is to gain an advantage. Quadruped playfulness, mock fighting, and bluffing, as in bear cubs and fox den mates, are *as if* situations with a training function. Sportive exchange is so typical of the animal kingdom that Milton gives it prominence in the most complete myth of origins in English elaborating on *Genesis*. In Eden fangs aren't yet aggressive weapons and cold weather doesn't yet account for fur coats:

> About them frisking play'd
> All Beasts of th' Earth, since wild, and of all chase
> In Wood or Wilderness, Forest or Den;
> Sporting the Lion ramp'd, and in his paw
> Dandl'd the Kid; Bears, Tigers, Ounces, Pards
> Gamboll'd before them, th' unwieldy Elephant
> To make them mirth us'd all his might, and wreath'd
> His Lithe Proboscis. (4.340-347)

The flexible trunk of the elephant is there to entertain fellow beasts, not solely to pluck high foliage. That animals are inquisitive parallels the more ambitious curiosity of Adam and Eve, which is prominent enough to draw a warning from Raphael. Linked to knowledge and station, curiosity is a prominent feature of the animal kingdom related to deception and to Satan's desire to find what it would be like to sit on God's throne. The satanic experiment is a rebellious inversion of divine decrees and an exaggeration of the egoism Adam and Eve are warned not to take too far. In line with the observation that creditable communication comes first and its perversion follows, Milton has the anti word, the archetypal deceiver, spring forth the moment the Son is made the Word, whose function is to enlarge what is communicable of the infinite Father to the angel host. Eventually that is the answer to evil's response to good even for humans, who are raised from a terrestrial to a celestial paradise with a correspondingly enlarged sense of the divine source.

In natural history where creatures are guileful it is sometimes in making as if they aren't around or in actually going into hiding, as Satan and his crew retreat to a remote corner of Heaven to conspire. Both hiding and trickery are as old as predation. Once cephalopods became predators they developed a sense of conscious contact as Peter Godfrey-Smith's tells of cuttlefish and the octopus. The stories he collects from personal experience and from other researchers are surprising in connections between the cephalopod subjects and their handlers that reveal mind reading. Hiding *in* ambush characterizes many predators and hiding *from* ambush their prey. Trying not to be detected presupposes knowledge of predators, which is why gophers, moles, pika, chipmunks, voles, mice, rabbits, and shrews hide away. If they lacked that deviousness, foxes, coyotes, and owls would make swift work of them. Pretending to be dead, a variant of hiding, isn't limited to pestered opossums. Pets going under the sofa at pill-taking time are applying one of the oldest self-protective devices of the animal kingdom. The skitterish fish, the scurrying lizard, and the color-changing chameleon survive by means of it, fish often in unison.

Imagining an intended injury can trigger a response as readily as detecting a real intent. Again Milton's myth of origins offers insights. Satan sees the Son's elevation as a slight and a damaging reduction of his own status. Given God's foresight, he may be right. Every fall including mankind's is foreseen, but not every subordinate being realizes that and hence thinks that concealment and deceit offer possibilities of escape. When damage comes of foreseen weakness in the creature, the provocative first step in a sense collaborates in the result. Milton clearly doesn't want the text read in that way, but that is the quandary of using the *omni* terminology. Everything everywhere is foreseen by an omnipotence that could have prevented its undesirable episodes.

Playing either dumb or dead is of a piece with the human role playing that Erving Goffman (1959) finds in image management. At masked balls, social groups make a game of it. On stage and in movies, actors make a living from it. In everyday life adults use it to project personas to fit social occasions. Where eye contact issues a challenge, purposefully not seeing is a playing-dumb variant. It disengages the pretender from something he would prefer to see go away. Human ancestry long ago learned to feign ignorance. Selective vision and hearing, which begins at the level of one animal ignoring another, rises to not seeing holocausts and the collateral damage of warfare. Neither what that has come to nor where it came from is all that puzzling when it is in so many places, at so many levels. It is as instinctive as spying and hiding in predators and prey. It is for a reason that bent cops and legislators who turn a blind eye to corruption have become plotting commonplaces in television, books crime thrillers, and movies. Both hideaway and predator roles can be put on hold when prey and predator declare a water-hole truce and have a life-sustaining drink. Calling a time out or declaring a truce is as necessary among competitors as the suspension of hostility among warriors. Every party can then come out into the open for a time without being defensive or being tempted by aggression. Setting a boundary to a field of play is one of the more sensible things competitors do. On the field, participants assume assigned roles. Off the field they can relax or follow biology to the dinner table. Calling time out is one of several peace-making devices that heads off the each-against-every embattlement Hobbes assumes to be the state of nature.

Word-Assisted Duplicity

The first articulated sounds broken into words are as lost as bird calls whistled on the wind, but that doesn't make one speculation about the deceptive use of words as valid as any other. It is likewise impossible to say when spectral creatures first entered the imagination, but if we include mistaken impressions and things that go bump in the night, they would have preceded full-blown language by millions of years. Given the mistaken sounds and sights that spook animals, underwater as well as land ones, that is undeniable. Spooking is a primitive nerve reaction and fake spooking basic to playfulness. Dimly imagined forces would certainly have predated Lucy and Ardi in the savannah days of *Australopithecus afarensis* and *Homo erectus*. The first and most basic level of illusions is uncertain perception teamed with caution and fear, and these too are general in the animal kingdom. That comes of real dangers and the advantage fast recognition and reaction have over sluggishness. There again what is ancient is also current, and what is contrived and false builds on what is real.

Brain size and bipedal locomotion are two other factors that bear on communication and trickery. Richard Klein and Blake Edgar in *The Dawn of Human Culture* (2002) opt for neural development as the catalyst of a cultural surge some 50,000 to 40,000 years ago, often associated with the development of grammatical language. They do so on the grounds that the anatomy had long since reached human proportions. Looking more to information management and words as "strange attractors," cognitive linguist Robert Logan (2007) believes that once language reached a certain stage of development it served as the catalyst of its own growth. That is a variant of the feedback loop principle. With innumerable combinations of phonemes available, words would have accelerated the organizing of concepts through tiered abstractions, and concepts in turn then generated more language.

Where a group collaborates in deception, there may be no effective check from within except whistle blowing. The zoological variant is discipline administered by a parent or senior member of a group. Several modern cases have been well publicized. Expand cheating and whistle blowing to the level of nations, give the whistle blowers computer access to thousands, even millions, of documents, and you have the Edward Snowden, Wikileaks, and Panama finance tangles. Donald Trump's habitual misstatements are deceit hidden in plain sight. Private enclaves of intelligence agencies and government bureaus sometimes find themselves exposed before the court of public opinion. Collateral damage inflicted in military attacks is caught on film, the footage gets released, and what an inner group considers necessary the general public calls murder. Had Hitler and company won War II, the atrocities of the Holocaust would have been either justified as necessary damage or denied, as they still are by Nazi sympathizers and other anti Semitic enclaves. Resistance to deception from inside the conspiracy may be vocal from the outset or develop secretly.

Legends, gargoyle-like demons, gnomes, and archetypes awaited not only extended narrative but systematic doctrine and abstractions. Whether extended myths first came from Anatolia, Mesopotamia, or Egypt isn't important in this context. Following clannish chiefdoms and roughly democratic councils in groups of limited numbers, they came in league with the deified kings of city states. If squabbling capuchin farmers and extant primitive tribes are any indication, territoriality and property had much to do with that and with hostility, human predation, and trickery. Land was a prime factor in appeals to high myth powers operating on behalf of kingdoms. Bows and arrows extended the range of hurled missiles that had already extended the range of the arm. Shorter range goes with smaller groups, longer range with numerous combatants and larger territories.

The greater the range and more powerful the engines of war, the greater the number of combatants and the higher the calling needed to assemble and command them. Big armies, big causes, more eminent leaders, more amplified rhetoric, and bigger gods, the latter because the difference lies not only in weaponry and but in the coordinating of forces. Mutually reinforcing illusions and group identity form a feedback loop as one intensifies and perpetuates the other. That could not have had a parallel in pre language behavior or in an animal kingdom with at most a few hundred distinguishable calls. Imagined places need imaginary powers to create them, and these need names and stories to put them into operation.

Putting Up a Front

Manners and role playing have no decisive separation. Their proportions vary from person to person, even moment to moment. Speech acts make use of both. Irony, sarcasm, and mockery, for instance, say the opposite of what is meant in voices that indicate the rules in play. Used in the wrong places they are unmannerly. In the right places and wittily they are entertaining. Group assigned identities look like make-believe on stage, but they have a way of becoming half genuine as indoctrination takes hold and shapes the thought and behavior of masses. As Philip Zimbardo (2007) has demonstrated in controlled experiments that resemble game playing, subjects take their role assignments seriously even when they know they are participants in an experiment. In their own minds, play prisoners over time become real prisoners, play guards real guards. Adopting different identities is common and not exclusively human. Among primates and quadrupeds, a new alpha male or alpha female takes on the role vigorously and a deposed one assumes a new status despondently. Some deception is likely to accompany role playing in primates. Any posturing is deception in the sense that it buries other sides of a more varied or more habitual character.

 The alternatives for animals with predators in the vicinity aren't exclusively flight or fight. Deception and bluffing, forms of role playing, can also be effective. As an enacted mini story sequence, fake an injury to lead a predator away from a nest or from offspring, as birds do, then bolt for cover. Put up a ferocious bluff, or stay unobtrusive. Meetings around conference tables put the equivalents into words and gestures. Naval warfare in the sailing ship era developed a sizeable repertoire of misleading indicators of identity and intent. Armies did something similar on land. Group identity assigned by conventions of that kind fades by degrees into individual pretended identity all the way back to early predators. The fronts that animals invent come in much simpler forms, of course, such as feinting, hissing, rattling, and flashing teeth. These select a defensive front from a narrow range of behavioral choices. Where a rival calls a bluff the dramatized role has to be played out. The line between real and illusory role playing isn't very decisive at any level of intelligence and sophistication. Once social and civil orders grew large enough and abandoned council leadership, the role playing of the elite adopted higher levels of pretense. Several millennia of pomp and glory in demi-god emperors, kings, and their courts testify to that. Mesopotamian dynasties used a terrifying front as a matter of course, broadcasting their atrocities in order to intimidate besieged cities and other threatened parties. According to scripture, Sennacherib dispatched delegates to Jerusalem during the reign of Hezekiah to taunt its defenders before attacking. The boasting continued in the inscription later made in his memory. Assurnasirpal II followed Assyrian practice in broadcasting what would happen to those who resisted: noses, hands, and ears cut off, eyes gouged out, cities razed and burned. He might have been exaggerating, but he wasn't entirely bluffing.

As pervasive as hierarchies and competition are, they don't necessarily generate adversarial situations like that. Though warrior role-playing among males ancient and modern is widespread, it isn't universal. Warlike tribes like the Yanomamos in the rainforests of Ecuador and Brazil may have been less pervasive among stone age people than anthropologists once supposed, though that remains a debatable point: "The bulk of ethnographic descriptions on record today," Christopher Boehm (1999) maintains in *Hierarchy in the Forest*, "are of tribal societies whose egalitarianism extends back… into the Paleolithic [early stone age] era" (91). Ferguson, Rousseau, Morgan, and Engels also take that position.

The noise that aggressors raise might be a coverup for fear at times, but it is designed to instill terror in potential victims and boost confidence in united aggressors. Prey that flees are vulnerable, because their defenses mostly point forward, kicking zebras excepted. Flight as opposed to an organized retreat becomes individual. A wolf pack aims to get a caribou or moose to run. Half a dozen that turned to fight would stand a better chance, but that's not in their makeup. In modern martial arts exhibitions, contestants accompany the physical role with noise-making. That is as much coached as it is natural, but it occurs in unsupervised child's play as well. In non stealth group hunting, our midrange ancestry, meaning a few million years ago–assuming again with Tudge that it resembled that of chimpanzees–became aroused as a group on the attack, swarming, howling, gesturing, and raising adrenalin in a collective furor that multiplied arms and teeth. Gangs in the California gold fields were made up of family men who were probably peace abiding in their former lives and again later, but their attacks on Chinese and Indians followed another psychology. Rebel yells, blood-curdling Indian calls, bombers in squadrons, and screeching mobs coordinate and hearten fighting groups.

In preparation, marching music and battle hymns of the republic are synchronizing rehearsal forms that draw participants into regimented patterns. That too makes use of groupthink illusions. Goose stepping, mass-marching Nazis didn't mind being filmed. Scripted by propaganda-instilled illusions, they were bolstering their morale and intimidating their victims. Football in the US puts on a militarist front, on the playing field in formation attack and defense and in opening ceremonies and halftime in flag adorned ceremony and marching bands. These maneuvers are reflected in stadium-effect synchronized cheers. The paraphernalia surrounding the game nationalizes the teams and audiences. That psychology turns out to have been pre battle preparation when real war breaks out against a rival 'team' abroad. Missiles, bombs, and bullets fill the air rather than missile shaped footballs. First propaganda to seed illusions in the popular mind, then shock-and-awe fury.

The first phase, already prominent in the political platform the youthful Hitler had adopted by 1920, carried over a long-standing doctrinal and cult bias from previous centuries, rephrased in Hitler in medical terminology: "For us, it is a problem of whether our nation can ever recover its health, whether the Jewish spirit can ever really be eradicated. Don't be misled into thinking you can fight a disease without killing the carrier, without destroying the bacillus. Don't think you can fight racial tuberculosis without taking care to rid the nation of the carrier of that racial tuberculosis. This Jewish contamination will not subside, this poisoning of the nation will not end, until the carrier himself, the Jew, has been banished from our midst," so a shorthand transcript of a speech at Salzburg in August of that year records (Koenigsberg, 2016). At that stage and at insane moments later, Hitler apparently believed his own myths and was encouraged to do so by the approval of his audiences. At each stage reason, discrimination among alternatives, and individual initiative got buried a little deeper. Once a game or a Ionesco play is over the spectators can leave the arena and return to something more normal. Once any great number of people have declared truth to be what a leader has declared it to be no such exit remains. The war is on, and in each skirmish the losers are dead.

It is only fair to mention that diplomacy and peace-making put up fronts as much as group-targeting propaganda does. Deceit and role playing aren't inherently aggressive but have many uses good and bad. When it comes to scouring history for examples we also need to guard against assuming that texts represent a given civilization as a whole. Populations and leaders can and do avoid the militancy and fanaticism of partisans grouped by illusions. Research from zoology to modern history finds mostly peaceful populations. Being newsworthy, turbulent times are more likely to be publicized than peaceful times. For every scam that makes the news are many honored contracts. Generally speaking, people earn the trust we place in them. The chief exceptions are volatile mobs and nations under the influence of stereotypes and partisan fervor gone viral.

Signs and Signals

Language studies like to point out that such things as broken twigs to someone tracking a bear are signs, not the signal a trail of bread crumbs would be. A dog barking out of excitement isn't being especially communicative, as one is that tells its provider it's time to eat or an intruder it's time to leave. There may seem to be a world of difference between an animal capable only of leaving signs and one that issues signals, but the distinction easily becomes blurred. Perhaps the most surprising finding of cephalopod research is the capacity of an octopus to issue signals to people. A warning from a watch-guard prairie dog is definitely a signal, sent and received. The basic difference parallels the one between the reflex camouflage of the chameleon and the calculated deception of the bird that fakes a broken wing. The latter presupposes a misled recipient whose mind is being read ahead of time and manipulated.

Some degree of mind reading is present in all communication, but deception implies a breach between sender and receiver, like a signpost at the crossroads pointing the wrong way. Whatever the intent, a believed deception puts the sender one up on the receiver. In pranks and practical jokes, the idea is to reveal that momentary superiority to the one just now tricked. In adversarial contests the benefit may be a dollar saved and for the tricked a dollar lost. Shared illusions are in a class by themselves, as a storm becomes the doing of a personification easier than it becomes a natural process that happens to spare a wheat field. "Thank you Lord" comes easier to some than "thank you jet stream and random luck." The unnatural substitution seems natural. Signs sent to no perceptible receiver play out as if the mind of the latter were readable, and many a seer and oracle has sworn to making requests that were granted. In both life and art, sage and serious people say they know what God wants. Ministries of faith put that in sect-defined terms and cite authorities in the far past. Congregations gather to join in giving ceremonial thanks, assuming that they will be heard. Grace said at a table does the same despite the food on the table being something the stars initially prepared in cooking up the elements, the soil grew, and someone in the kitchen finished cooking.

In all variants of role playing it pays to keep in mind the basic distinction between two kinds of simulation, openly fictional and intentional dissimulation. Strong attractions make good samples to scrutinize for masquerading. So is the courting of allies at the national level, since both adjust courtship protocol to the addressed parties. Courtship may be no more than flattery or it can be part of a campaign. In the animal kingdom, preening falls in that category, in that case not much more than putting the best foot forward. If a female reacts favorably to a suitor's dance-like excitement, the suitor is encouraged to repeat the performance, converting an accidental play-acting move into purposeful display, in effect into rhetoric. That move may then become a convention selected by other suitors. It enters the local culture. Uplifted, the courted party becomes a power animation, a god or goddess or perhaps an emperor declared to be a demi-god. Ronald Englefield in *Language: Its Origin and Relation to Thought* (1977) wonders if the initial move at the animal courtship level need even be intended. Whether it is or not, as females learn to distinguish among copycats, sophisticated variants follow.

As Kenneth Burke points out concerning Castiglione's *The Book of the Courtier*, courtship is a form of negotiation in monarchical situations. What works is repeated until it doesn't work, at which point a variant is introduced. Still another step in that progress, Englefield finds, is the abridgment of copycat signals, capitalizing on familiarity for brevity and energy efficiency. The addition of language is needed to initiate instructed practice and cultural indoctrination. Whether instinctive or learned, the roles are always partly make believe. Theydon a behavioral mask equivalent to a warrior-role enacted in uniform. A good deal of behavior is a put on, at the con level done for gain, in social roleplaying used to establish a hierarchy, the first a put up job, the second sometimes a put down. Put on, put up, put down: the emphasis is on performing artifice of codified acts. From the smiling or crying infant to the adult dressing to go out, personality is to some extent artifice.

Artful simulations such as cave drawings were an early phase of iconic records. Pantomimes, too, could conceivably have preceded grammatical language. From early civilizations onward, social orders used headgear, robes, movement, and symbolic objects such as tridents and globes to indicate rank and office. These were indications of communal acceptance of the myths they enacted, as a crown set in place has a social agreement behind it. Clive Ponting presents a bronze and iron age phase of that story in *A Green History of the World* (1991), and Jared Diamond a comparable version in *Collapse* (2005). The costumes portrayed in the Merriam-Webster's *Encyclopedia of World Religions* (1999) represent a sampling of the world's current ritual enactments and art forms. The costumes are accompanied by mask-like faces with a stage production look. Wigs are piled high. Chest-length beards, elaborate ornaments, jewelry, and feathered headgear distinguish one office from another. Solemn monks sit in pre-formatted postures in the body language of piety.

The audiences of such displays are cultural initiates or cult followers coached in the interpretation of what they were seeing. The influence on them of costumes alone shouldn't be underestimated. Dress is an artifice that priests, kings, courts, and palace guards as well as warriors use to indicate their functions. When one office holder steps out of uniform another steps in. Both the offices and the supernatural figure behind them diminish as empiricism grows, a general trend in realistic secular social orders. Marginal, outcast, and outlaw groups follow assignments that mirror those of conventional equivalents. Conformity is the goal whether the effort is straightforward or devious. Coordinating a swindle creates something like a repertory player group with a script writer or director. The carryover from one collective to another is instinctive cooperation in a sociable animal. It can be applied to pirate gangs, armies, or the League of Conservation Voters. It doesn't matter. Good, bad, or indifferent, the cohesive element is much the same in nature if not in intensity. The stadium effect at a football game is part of a sport. Everyone knows it is a game. At a speech by Stalin it is deadly serious and more deeply infectious.

An appointed spin doctor for a political party (not that it resembles an outlaw gang) didn't become a recognized position until modern times, the origin of the term being American from the 1980s, but collaborating civil and religious orders that assume a public image have a long history. Public speaking depends on the status and authority of the speaker, and nothing comes higher than delivering messages from an exalted source. Stevens' "Of Heaven Considered As a Tomb" (in *Harmonium*, 1923) asks far seekers a question no longer likely to be answered, most recognized harbinger roles having been eliminated if not all self-appointed ones:

> What word have you, interpreters, of men
> Who in the tomb of heaven walk by night,
> The darkened ghosts of our old comedy?
> Do they believe they range the gusty cold,
> With lanterns borne aloft to light the way,
> Freemen of death, about and still about
> To find whatever it is they seek? Or does
> That burial, pillared up each day as porte
> And spiritous passage into nothingness,
> Foretell each night the one abysmal night.

Stevens accompanies that with a claim in "A High-Toned Old Christian Woman" that "Poetry is the supreme fiction," or in reverse, a supreme fiction is foremost poetry. That is in recognition of what has always been the case but not acknowledged as such with message bearers thought to be authentic. Demi-god kings and emperors had the highest possible sanctions that came in their case from mixed human and divine parentage. It remains a common custom to make ceremonies solemn by having them hearken to "the darkened ghosts of our old comedy." Important collective missions gain in dedication by means of just such intangible presences, as do individuals convinced they are responding to an actual calling. Thinking that certain messages come from far off was initially the product of a shrunken universe in which mankind was important, a master delusion in common among myths that disagreed on much else.

My critique has of course assumed that it is all to the good in natural philosophy that information now comes not from a winged Mercury or an archangel but from the opposite direction, from earth and sea, from animal studies, from the deep past of the Cambrian explosion, from rock evidence of the early solar system, from the sea dwellings of eight-arm creatures. The spiritous passage into nothingness is no less haunting for our taking it to be a metaphor, indeed more so with the raised power of exponents and with photons as message bearers. That contrast is almost as stark as the one between five-pointed paper stars and the gigantic nuclear furnaces to which we owe the elements and ourselves. An animation that takes over that enormous machinery trivializes it in inflating the importance of the inventive and sometimes delusional biped in existence only some $1/69,000,000^{th}$ of time elapsed since time 0.

Prosody

It may seem that I've strayed from skull duggery in including fabrications of whatever kind, but not really. The devices with which language comes equipped are the doing foremost of the self going forth. We can't say that it does so genuinely or falsely when nearly every speech act mixes the two. Hence I should say the concocted self puts in an Mlodinow appearance, since no one knows if an undivided, single self exists or we consist of accrued parts. Tailored roles come with assumed character, style, and emotion. Feeling attends every transcription of selected aspects of a complex self. The term *prosody* takes in the rhythm, pitch, tone, volume, and attitude of verbal staging. In keeping with the assumption that objectivity doesn't come easy, Steven Mithen proposes in *The Singing Neanderthals* (2006) that semantics would have been harder to master and come later than such prosody elements as intonation, pitch, and rhythm, the instruments of tone and feeling in proto as well as grammatical language.. Proto language and prosody bring a different satisfaction into play, as humming a tune does compared to voicing its words. In singing neanderthal terms, the social pleasure of holistic expression is akin to that of grooming (136). That is an attractive proposal so long as we add that the emotional range includes irritation and competition, which can be more like scratching than grooming. It includes anxiety, hope, and more. If gelada monkeys can use rhythmic and melodic utterance for solidarity *and* hostility, certainly early hominids could have done so prior to S-V-O syntax and semantics. Confusion too is a feeling, merely not a well organized one, as we can see in the fluttery irrelevance that breaks out in animals momentarily lacking purpose and direction.

Mostly, though, prosody is coordinating, as rhythm synchronizes marching feet. A west highland terrier of my acquaintance responded to operatic sopranos, preferably Violetta in *La Traviata*, with howling that went up and down in many of the right places but without rhythmic synchronization. To all appearances he was expressing pack sympathy. It is true that what he was thinking couldn't be deciphered exactly, but the response was so compelling to him that a simulated Parisian courtesan over a century ago took the place of fellow canines. To an informed listener, operatic prosody teams up with the semantic content of the story and *explained* feelings. To someone from a different language unaware of the story, the reception would be a step nearer that of the terrier. Music comes without language but language never without prosody. Voices go up and down and assume a pace. Weariness and depression are as dependent on that as pleasure and excitement. Their ups and downs are merely closer to a flat line.

If quadrupeds, troops of monkeys, geladas, and flocks of birds communicate emotion, we should have no difficulty imagining *Homo habilis* and *erectus* doing so long before grammatical language. As James R. Hurford speculates in "The Language Mosaic" (2003), the chief preparations for semantics were sounds and gestures, pre-syntactic in organization, pre-verbal in concept, and social or pragmatic in function (41). Thus when phonemic language did finally arrive, it found resources in place to build upon much as the alphabet later had significant vocalization in place and needed only coded ciphers to represent it. Assuming again that fakery followed where genuine expression led, the emotions of prosody joined other aspects of simulation. Concepts of language capacity and its evolution, especially those devised by linguists concerned with syntax and semantics, tend to be skimpy concerning such proto-language increments as tone, mood, rhythm, emphasis, and pace, which aren't merely appendages to meaning but essential to it. Cross species communications can work very effectively on merely the prosody level without the semantics. When we come upon a hissing, tail-rattling snake we have no difficulty knowing what it means. Trained dogs respond as well to hand gestures and whistling as to verbal commands, though in that case the signals are semantic imperatives. Irony, sarcasm, and other aspects of mockery and mimicry have prosody elements incorporating simulation and can be faked in non linguistic forms of role playing. A friendly gesture, a touch, conveys a connection between quite different levels of consciousness. Reaching out to pat the neck of a horse conveys something nameless to the horse. Directions from reins and feet convey something specific that could also be voiced.

Those who work with *poesis* are aware of all this, and those who study rhetoric are advised to become so if they aren't already. Hype gains in prominence with advertising, ideological partisanship, and harangue, and it is often signaled by a raised voice. Its opposite is understatement (litotes), which when combined with irony becomes an inverted form of hyperbole, as when Mark Twain remarks that reports of his death have been highly exaggerated. That twist in communication is unavailable to non human intelligence, but animal playfulness comes close. Can an octopus feel scorn? Probably not, but it can demonstrate distaste. Food rejection is a primitive form of what become more complicated acts of discarding. Getting rid of something painful or shocking is similar. Even a chart of incremental evolution as impressive as Jackendoff's, which includes primate conceptual structure, symbol usage, hierarchical structure, inflection, semantic relations, and grammar, leaves out much that will have to be included eventually. Certainly how the brain attaches affective accompaniment to sorting, attaching labels, and organizing warrants close attention. Enhancing communication and enhancing thought aren't particular enough for a field that includes competition, exertions of influence, trickery, and the purposeful disrupting of a rival's thoughts. Literary critics often become adept at accounting for cadenced prose, the rhyme and meter of poetry, and other aspects of prosody that bind feeling to meaning at advanced levels of communication. Music and art critics, too, connect formal elements to feeling in league with meaning.

That repetition and impassioned speech are more prominent on the podium and pulpit than in the living room underscores their public function and makes them a critical part of the study of rhetoric and on-stage dramatic speech. Rant heaps up prosodyin large amounts and selects evidence to give a calculated impression. Prosody committed to selling things including ideas fills the newspapers, magazines, billboards, internet, and television. As simulation, mainly of enthusiasm, it belongs under *poesis* but in intent is rhetoric and in style hyperbolic. In politics, prosody and rhetoric all but replace plain speaking. As we'll see later, epic and oracular literature stage similar high-prosody performances, epic in taunting, threatening, and rousing support, oracular literature often in turning feeling against foreign adversaries and the local half-hearted. Prominent in both epic and oracular sermon are the imperative mood, didactic instruction, and the cadenced presentation of rewards and punishments. Biblical jeremiads are similar to the incantations of *The Epic of Creation* and *Erra and Ishum* in offering impassioned lists and parallel enumeration devoted to recruiting enlistees in a holy war. These add little semantically but a great deal in emphasis. In battering down resistance, repetition has a persuasion value of its own added to pitch and volume. Rhythm is to forceful delivery what waves are to mechanics, the temporal organizing of force. Oft repeated imaginary things come to seem real. Many of the devices in Renaissance rhetorical manuals were taught to enhance the delivery of connotation along with denotation. They were extracted from classical sources of great age that in turn had undoubtedly codified prehistoric oral practice.

The better forms of eloquence keep prosody in balance with semantic elements and to some extent justify Plato's distinction between philosophic dialectic and sophistic rhetoric. It is true, as Jaques Derrida's analysis of "Plato's Pharmacy" in *Phaedrus* demonstrates, that neither Socrates nor Plato can avoid the pitfalls of rhetoric (can anyone?), but in its gradations of prosody, podium rhetoric is more loaded than the Socratic method. The formal speeches on display in literature make liberal use of heavy cadence and repetition. The public speech techniques of Greek assemblies and Roman forums that came to the English Renaissance filtered through Seneca and Cicero had much to do with the heft of Shakespearean speeches, complex combinations of attitude, feeling, hierarchy, and sometimes philosophy, the latter prominent in *Hamlet*.

Literature contemporary with the manuals in rhetoric such as Sidney's *Arcadia* and Lyly's highly mannered *Euphues* made elaborate use of them. Milton assisted the Puritan rebellion with the combined help of biblical eloquence and Ciceronian oratory. In *Paradise Lost*, too, the formal delivery of speeches as if from a podium is a prominent device in Heaven and Hell if not in Eden, which has personal instruction but no oratory except in Satan's address to the forbidden tree, staged for Eve's benefit. Milton sets that particular speech up as an example of what will later come to pass in the rhetoric of Athens and Rome, sometimes as manipulative and deceitful as the first human exposure to a masterful deception:

> With show of Zeal and Love
> To Man, and indignation at his wrong,
> New part puts on, and as to passion mov'd,
> Fluctuates disurb'd, yet comely, and in act
> Rais'd, as of some great matter to begin.
> As when of old some Orator renown'd
> In *Athens* or free *Rome*, where Eloquence

> Flourished, to some great cause addrest, Stood in himself
> collected, while each part,
> Motion, each act won audience ere the tongue. (9.665-674)

What follows is an address to the tree as if to a god, entirely for show:

> O Sacred, Wise, and Wisdom-giving Plant,
> Mother of Science, Now I feel thy Power
> Within me clear, not only to discern
> Things in thir Causes, but to trace the ways
> Of highest Agents. (9.679-683.

And so on with an imaginary power that ought to have collapsed of its own weight with the three hefty adjectives leading up only to "Plant." Assigning the power of enlightenment to a tree rather than to its creator is an error of importance in a hierarchical creation. To violate any command from this source is to throw the created universe into jeopardy. Satan's success is with only a segment of it, but that segment happens to include all human history.

Poetry concentrates prosody more than prose ordinarily does. Consider a couplet from Alexander Pope's "Rape of the Lock": "The hungry judges soon the sentence sign,/ And wretches hang that jurymen may dine" (3.21-22). With equipoise and cadence to time the delivery, Pope transforms the callous satisfaction of the judges' verdict into a judgement of the judges. In coordinating epigrammatic wit and metrical units, the heroic couplet often uses a midline pause that breaks two lines into four units. These serve as balance and delay for maximum impact. Rhythm, rhyme, and emotional satisfaction, even in a mock form as polished as this, aren't merely vestigial. They are inseparable from the phonemic elements that deliver the meaning. A sentence with or without much prosody becomes a number of things besides an integral grammatical unit—a sculpted feeling, a formal presentation, a performance. In writing, a capitalized opening letter and concluding period package its meaning. That framework is similar in contrivance to a framed painting on exhibit or a sonata except that it specifies meaning more exactly. Semantics pinpoints where the feeling is to land where music without words does not. Listeners may unwrap sentences bit by bit, but with the final word the whole clicks into place as a structured unit in an imperative, exclamatory, interrogatory, or indicative mood. In Northrop Frye's distinction, *dianoia* or meaning is *statically* present throughout. It is as if a plot were a sentence writ large, each part in league with the others. The final effect is almost as statuesque as it is musical. In verse, metrical and rhymed units click into place like the locking pieces of an arch. Regimes have been toppled with the help of chanted slogans and rhymes. "When Adam dolve and Eve span/ Who was *then* the gentleman?" the Lollard priest John Ball asked in the Peasant Rebellion of 1381. He was hanged, drawn, and quartered himself, but the catchy rhyme lived on. Liberation from serfdom eventually came with the help of similar drumbeat pulpit and podium rhetoric.

As we've seen, whatever adds point to communication can also be used on behalf of delusions to make them infectious. Though Ball and his contemporaries were wrong about Adam farming and Eve weaving garments, the point with respect to social hierarchy was valid, as was the 17th century conflict between cavaliers and middle and lower class Puritans. With the help of prosody an illusion takes on rallying force. As a rhetorical intensifier, religious doctrine could be used by both sides of a social/economic conflict. Having nothing tangible behind it, a ghostly presence is ideal for verbal inflation. Semantics provides the specifics and the direction, prosody the fuel and energy. Which came first in the development of communication and its skullduggery branch may never be known with certainty, but we do know that quadrupeds as well as bipeds have feelings and means of expressing them. It is harder to imagine creatures of the deep other than the mammals having them. A fish can be hurting but seemingly can't have hurt feelings. The evolutionary sequence probably went from nerve and brain development to consciousness and so on eventually to self consciousness, leaving plants, worms, and fish behind. The desire to communicate what the self sees undoubtedly pushed the development of a means capable of specifying both objective reference and subjective response, the most urgent and elemental branch of which would have been the *watch out!* of the parent or watch guard. Coming along somewhere in the rear where irony and humor mock leading role players is the falsified equivalent calling out *Wolf!* when none was there.

Chapter Five

From Myth to Philosophy and Science

Words without Things

Until something grows exceptional, mythopoeic imagination isn't normally concerned with it, although magic in medical situations descends to common plants and chemicals. Such things as plants, insects, animals, and weather usually have to be extraordinary to attract tall tales, either extraordinary or simply prominent like the sun and moon. Some of the transitional phases from predominantly mythic to scientific and philosophical principles we find in Greek literature and philosophy from Homer and Hesiod to the pre Socratics, Plato, the historian Thucydides, and Aristotle. The former two were poets with prominent meddling gods and goddesses that the readers and most likely the poets themselves believed existed. In contrast, analytic thought in the late 5th century started applying topical breakdowns rather than storied personas to phenomena, a practice carried out with some thoroughness by Aristotle (384–322). Historical events produced theses that were systematic enough to be outlined with headings and sub topics. Except for their influence on plebs, demos, nobles, and slaves, mythic heroes and deities scarcely matter in Thucydides and Aristotle. Math in Egypt and Greece, astronomy in several places, and chemistry in Egypt and Greece were well underway by then.

Let us first retrace enough of last chapter to be aware of what likely came with grammatical language or at least couldn't have preceded it. It isn't that unusual to imagine going without words even now, but it takes an effort to put ourselves in any strictly pre linguistic situation. Among those whom Gulliver encounters at the Academy of Lagado are three professors in the school of languages who abolish words to save their breath. They carry actual objects about and point to them. From those born without the capacity to speak and hear we know something about what that is like for creatures of intelligence with no such objects at hand. In domesticated animals that understand some words but can't speak, we can see how a lesser intelligence deals with limited communication, usually by means of inquiring looks and responses to gestures and to a few words. For the period of development after the current human brain size with limited sounds under command it is nonetheless not easy to reconstruct exactly what it must have been like to have an urge to communicate without adequate means to do so.

Suppose it is 60,000 years ago, before the Middle/Upper Paleolithic transition, somewhere in what is now southeastern Europe. Like other creatures, I like to identify what comes before me. I come upon a lake ringed with golden foliage brilliant in the sun and am struck by the spectacle but have no names for backlit fall colors. The sun, the sparkling water, and the leaves seem all of a piece. Had I the words, wonder, beauty, and something spiritual might be applied to the scene, but poetry hasn't been invented yet, and I am locked in silence. Suppose further that along a coastal bluff in the spring I see a comical bird, a tufted puffin, digging into sandy soil. Others of its kind are adding dead grass and feathers to their burrows. Putting like things together I associate it with other burrowers and nesters. Nesting makes me think of eggs and chicks and growth to maturity. A different bird flies along the shore and dives into the water for fish.

Curiosity over feathered life makes me want to know more about such creatures, what they are, what they eat, how they behave, but the terminology of species hasn't arisen yet either, and again I am left in silence. A larger bird with a hooked beak and white patch above its tail, a harrier, gliding low over a marsh reminds me of hunting and how tribesmen hunt. The marsh hawk hunts alone. We coordinate our stalking with muted calls and gestures. The urge to connect further, to teach and to learn, is strong, but we lack any great amount of pedagogy, and when we are out of one another's sight and hearing we've no way to communicate except for monosyllable shouts and a few scratches on the ground.

In basing accelerated learning on the need to organize and convey perceptions and information, Robert K. Logan (2007) in *The Extended Mind* considers what must have been the psychological pressure to invent grammatical language. Each phase from speech to writing and on to mathematics, science, and the internet has come, he believes, in response to information overload (41). As my hypothetical example suggests, what applied the pressure to communicate wouldn't have been merely information but aesthetic feeling, a glimpse of connections, bonds among fellow tribesmen, and something like an urge to call out Eureka! at a discovery. By itself an impulse to speak wouldn't create syntax and semantics, but any fragment of language, any distinctive sound, would help. At the sight of the lake and golden foliage, "Ah!" with a gesture would go some way toward what someone equipped with language might say to a companion. Sensory-motor hookups were in place in animal life that bordered on consciousness long before anything we could call self consciousness. Both consciousness of self and names for objects would be necessary before subject-object relations could be put into grammatical relation. Syntax is the language simulation of actual subject-object relations.

When grammatical units did eventually materialize, a prolonged stagnation ended, so Derek Bickerton believes (2003): "Then, suddenly, creativity blossomed" (91). I don't see how it could have been otherwise, language offered such an unprecedented conceptual development. It might not have been all that suddenly, however, if the pieces fell into place one at a time. A day or two or a millennium could have separated one phase from another. Primate brains had long before developed a neocortex to complete the hindbrain, midbrain and forebrain and the left and right divisions. When the brain and the throat apparatus were ready, it would have been sensory-motor, cognitive, social, and psychological prompting that exploited them. The psychological impulse would have come partly from perceptions of likeness and of cause and effect sequences. The poetic and the scientific would have overlapped.

After the evolution of sentences, stories segmented into episodes as cause and effect sequences must have come in due course hooking grammatical units together. The first ones to craft artful anecdotes were explorers as surely as those who ventured into strange lands. They were as necessary to the sequential logic of science as they were to myth making. The equations of Newton and Einstein and tracking the filaments of cosmic superclusters owe as much to them as to cosmologists and the physicists who gathered yearly at the Solvay conferences. Whenever it came, in whatever segments, connecting sequences was a remarkable achievement. The timing would have been long before inscriptions on clay tablets and stone in the 4th millennium. Unrecorded myths of origin would probably have been connected to clan and tribal histories on the way to the gods of territorial states and empires on record. The verbal and physical tools, accumulated information, and human clusters in clans, tribes, and hamlets would have developed in conjunction, and in keeping with last chapter included the initial infectious fictions and illusions. We can't assume that illusions were chained to information every step of the way, but names for what didn't exist would have been abundant before any of them got inscribed on durable materials. The first god lists were extensive and the first recorded myths and epics well developed works of art.

As any concept of the progress from myth and other tall tales to philosophy needs to concede, these quite different ways of thinking, imaginative and rational, continue to alter in balance. Although they don't rate so much as a mention in Gilbert Ryle's classic *The Concept of Mind* (1949), made up things and sequential or connected observation have attracted appropriate attention in anthropologists concerned with mind development and myth studies: "Mythic culture tended rapidly toward the integration of knowledge," Merlin Donald remarks in *Origins of the Human Mind* (1991). The "repertoire of mimetic culture," in his terminology, had to come first "under the governance of integrative myth." That "signaled the first attempts at symbolic models… and the first attempts at coherent historical reconstruction of the past" (267). Myth thereby served a vital purpose on the way to "theoretic governance," vital to the organizing of information in sequences and categories. That expands on what Andrew Lang proposed some time ago in *Myth, Ritual and Religion* (1887) and John A. Wilson puts in terms of the cosmos and state in early Mesopotamian culture (Frankfort, 1949). Discourse studies also go over much of that territory in sorting out psychological differences between mythopoeic narration and the causal joining of one thing to another. Categorizing is an act of verbal domestication.

Fantasy, dreams, and daydreams put additional kinks in the already crooked path from myth to philosophy and science. If we want to understand animals and objects as they are we almost have to train ourselves *not* to personify them. Knowledge of basic laws of nature were still 5000 years away when the first personifications were recorded in writing. David Leeming's *Oxford Companion to World Mythology* (2005) lists over sixty geographic and cultural areas of myth based on animations. Gods and goddesses came in flocks. Anu, Enlil, and Enki, followed by Tiamat, Marduk, and others were among the noteworthy Mesopotamian ones parallel to Egyptian ones. The rise of human hierarchies in city states too large for democratic assemblies may have suggested that the gods, too, should be ranked. In any case it was undoubtedly human belligerence that made some of them war gods and figures of tumult. Gods weren't the source of storms. Storms were the source of stormy gods. Natural phenomena occasioned similar figures of the sun in scattered regions, the Semitic Hadad and Elohim, the Canaan Baal, the Babylonian Anu, and the Egyptian Horus and Re.

These and other personifications came easily enough, since they heightened what was familiar, as human wisdom in monotheism was heightened into all knowing, power into omnipotence, presence omnipresence, and agency supernatural agency. It is also true that attributing thunder to Enlil or Zeus was easier and probably more interesting to most than studying lightning or trying to clock the speeds of light and sound. That one way of thinking, the primitive, may have been necessary to the other, the analytic, is the opinion of some anthropologists, as anecdotal sequence was to philosophic cause-and-effect logic. I think it more likely that less systematic daily realism, the sorting of objects, and mechanical procedures graduated into more methodical equivalents, that explanations of limited scope grew by accretion into comprehensive ones. That the brain remains superstitious and prone to illusions in that raising of things known to higher levels isn't surprising. It retains much of its earlier (limbic) phases. When a perception comes with a name, say a bolt of lightning striking nearby, the name and understanding of the phenomenon are cerebral. The flash of accompanying awe and fear are as primitive as small fish fleeing large ones.

That the singular figure of Hebraic monotheism caught on and spread to Christianity and Islam suggests an additional psychological matter: the desire for an all-inclusive narrative and unity in diversity, the aim still of a Grand Unified Theory (GUT). The advantage that a city or territorial state gained from a unifying power may have figured in its projection of such a force. One major difference between the one that established a covenant with a nation, Yahweh, and his predecessors, as Benjamin Uffenheimer points out, is that the predecessors came with life stories. They had mates and offspring. With the exception of a lingering connection with the consort Asherah, nobiography comes with Jehovah (Eisenstadt, 137), and in a later variant only one offspring, the Son born of a virgin mother. That didn't make the Hebrew version any less a projection of human traits than Baal, El, Chemosh, and Asherah. Once the Deuteronomist history was assembled as a national story, the Israelites set about purging idolatry internally and from the Mediterranean area generally. Perhaps because scattered 'high places' remained vulnerable to outside influences and weakened the authority of Jerusalem and its kings, reform movements like that of Josiah (641–609 BCE) discouraged them. One advantage of a single figure would be lost if it didn't contribute to national unity. To do so it had to resemble a patriarch somewhat less, be more austere, and apply itself to the welfare of the cult and nation. The demand for absolute faithfulness in any given cult indicates the recognition among its leaders that clearing away tainted concepts is a prolonged process, one reason for new prophets arising and calling for the elimination of rivals. In the Hebrew texts genocide is the preferred method. In a worldwide evangelical mission that removal of rivals isn't possible, and so a separate place to put non cult members awaits an afterlife. To the embarrassment of Christian commentaries, monotheism in its origins made the Hebrew anthology a book of wars as well as personal anecdotes, axiomatic passages of wisdom, and stories of defection and treachery. The belligerence became a hindrance to later cult recruitment. The main point with respect to such a universal figure, however, is that it requires unquestioning conformity. In a strictly territorial kingdom it never gets much beyond the sponsoring nation's borders. In a non territorial new kingdom it goes international and its followers become evangelical. Their institutionalized doctrine looked systematic and was in its own way, hence parallel to philosophic and scientific procedure. Unlike natural laws, however, it proceeded by morality defined foremost as obedience to what the patriarchy had received, codified, and ritualized. Rival claims to that eventually divided the nation and set it against neighboring nations.

Philosophy and Disputation

Questioning what is real and what illusory seems to afflict most philosophers sooner or later from Zeno thinking motion unreal to Plato's cave of shadows. In non material theological variants of cause and effect what we can call the age of Satan, the arch deceiver, became one way to explain why so much had gone wrong with the world. Philosophy sometimes combined gods and demons in that respect with topical breakdown and systematic analysis culture by culture, something that Hesiod and Herodotus illustrate in quite different ways. We can attribute more empirical philosophic argument initially to academies where logical argument was highly valued. In 5th century Greece the art of disputation and issue definition developed across from rhetoric. Hesiod's blend of myth and history in *Theogony* and in *Works and Days* (2004) had put explanations of natural phenomena in animated terms, Earth (Gaia), sky (Ouranous), Chaos, Eros, Ocean (Okeanos), sun (Apollon), night (Nyx), darkness (Erebos), and day (Hemera). Abstract matters such as justice found minimal representation, as judgments had in earlier mythology. *Works and Days* devotes an extended passage to Zeus's quixotic judgments, typical of divine caprice that makes personifications fit natural history's indifference to right and wrong and human history's inconsistent attention to justice. Random luck in nature translates into captious temperament in animations. If language evolution were put in an evolutionary tree or bush diagram, formal disputation and the nomenclature of abstractions would warrant a branch of its own. Discourse in Aristotle is divided into disciplinary areas under a classification pyramid, with sub topics and particulars going down and graduated abstraction up. Theology and myths of origin are eliminated for the time being, though both Plato and Aristotle are incorporated into dominantly theological argument later, Plato prominently in Philo, Paul, Augustine, and Renaissance neoplatonism.

The thematic content of stories lodges somewhere in between myth and analytic classification, and it, too, can raise questions about phantom presences and marvels. The story teller can either pause to draw a moral or cause it to emerge in details. However it comes about, a theme converts a narrative sequence into what Aristotle calls *dianoia* or thought, which can veer away from concrete situations into speculation. As we'll see in a moment, Philo and Christian allegory developed a concept of layers that anchored biblical incidents in metaphysics and morals. Chronicles and fictions in story form thereby yielded articles of belief, sometimes under the administration of clerics and theologians, as a sermon extracts moral admonitions from a good Samaritan.

In the most formalized version of that moralizing, three stages followed an episode assumed to be historical, the second layer a later historical or narrative parallel offering a clearer revelation, the third a moral principle, and the fourth a glimpse of a higher reality. Thus something David did might prefigure something concerning Jesus. Both could then be moralized and offer a glimpse of the holy kingdom, the ultimate goal and the end of natural history as earth illustrates it. That systematizing of exegesis had the advantage of making the new dispensation more enlightened than anything previous, but even after allegory had done its best, historical revelations remained incomplete and thus called for faith and hope. That crystallized in cult recruitment side of the method and the admonition 'believe and ye shall be saved'. It winds its way through real and simulated chronicle history, works into moral advice, and establishes a higher realm as a cure for what has gone awry in history. Again as in Plato the nebulous is real and the real nebulous. A good deal of the western philosophical tradition is devoted to making that sound plausible and imagining the eventual elimination of the known universe.

Most conversions of narrative episodes into philosophy are less systematic than that. Theme, hypothesis, and theory provide common organizing schemes, but in some anthropology, before theoretic reason and philosophy could take root, myths had to go through personification in something of the way drawing a moral from an episode must first go through the episode. Theoretic governance can be equated to Ryle's "higher order acts," which require proactive minds and complex consciousness, as when a child, "having been separately victim and author of jokes" finds "how to play both roles" (193). Consciousness includes self consciousness and talking-to-oneself, which humans never outgrow and no other animal ever grows into. In Northrop Frye's *Fables of Identity*, James Frazer's *The Golden Bough*, and in Cassirer and his student Susanne Langer, personifications find common ground with philosophy and psychology in attaching general principles to proper names with a character and area of authority. Figures bearing different names take similar patterns in folklore, fairy tale, and romance as catalogued in encyclopedic guides like the Merriam-Webster *Encyclopedia of World Religions* (1999) and Ashliman's *Folk and Fairy Tales* (2004). Ritual and initiation ceremonies are much alike despite differing by sect and culture. In pre Socratic myth, Hesiod's *Works and Days*, the Hebrew anthology, and Homer's *Iliad* prepared the way for Hacataeus' *Genealogies* and Herodotus' *Historie*. Mixtures of thematic rationality and myth headed toward the more methodical rationality of Aristotle and Thucydides. Thucydides' highly factual *Peloponnesian War* extracted principles from examples without personifications. War councils, not Athena, steer the way into and through the war.

As I suggested earlier, whatever the preparation for methodical reason in fables and myths we shouldn't overlook the preparation that comes built into pragmatism. It has daily applications of method everywhere simply because not to apply reason and method to food, shelter, social relations, and health reduces the chances of survival. It is easier to upgrade pragmatism in practice to pragmatism as philosophy than to convert fables into applicable lessons. It is also true that thanks more to common sense than to science or philosophy, more diversity existed in ancient cultures than their umbrella myths and national identities suggest. National titles covered subsidiaries and regional groups not well represented in the literature. The subduing of rebellions claimed in Mesopotamian inscriptions and admonitions from prophets in the Hebrew anthology tell us that. For reasons Benedict Anderson spells out in *Imagined Communities* (1983), a national title generalizes. Niels Peter Lemche in *The Old Testament between Theology and History* (2008) extends the unreliability of the national Hebrew narrative in that and other regards through Kings and Chronicles into the book of *Ezra*. In his estimate we know very little of the 6th century Babylonian exile, and in *Ezra* Hebrew scribes were still inventing "a hero of Jewish faith" to recite "the law of Moses to the people" and thus complete "an ethnic cleansing of the Jewish nation" (Lemche, 160). The Dead Sea scrolls continue along that line into the second and first centuries (BCE).

Philo: To Disclose Hidden Doctrine, Allegorize

The attempt to squeeze meaning out of natural phenomena and make events serve a teaching function can become quite ingenious, as art and chronicle combine in a historical novel. As different as they are, philosophy, history, and fable collaborate in thematically guided narrative. Exegesis fills in where themes are only implied. The medieval Book of Nature became a venerable tradition in wholesale conversions of natural phenomena into virtual texts. Making the conversions was a prime intellectual occupation for centuries. Parallels between a supreme creator and human authors were worked into both natural phenomena and chronicle events. Any part of a creation commanded initially by the Word could conceivably be turned back into a revelation of its source so long as everything harsh is explained some other way. Another factor enters the deciphering of the text. Its author has kept secrets from the reader until the best moment to reveal them. In a sacred book where a contributing author is a god, the secrets are kept a good long while and released one at a time only to chosen parties. Prophets declare them and critics like Philo explain them.

Philo produced the first full scale elaboration of that principle in lifting an anthology miscellany into focused instruction. That set up over a millennium of explanatory method that attempted to explain away the obvious fact that mountainous, oceanic, and desert earth is harsh and unforgiving, that every creature is born or hatched on it with a mortgage on its life. Predation must somehow have passed from devils to the animal kingdom. That people are to blame can be seen at work immediately after the fall in *Paradise Lost*, though to make it actually come about requires a wrathful deity with more power than Satan. Chronicle incidents in Philo's handling come with built-in lessons that only have to be pulled free of extraneous details. The example he set was adopted by the multi leveled Christian exegesis where it had the final uplift of an anagogic level reserved for the new dispensation. The narrative plus the commentary produced a hybrid that was mythic, eschatological, metaphysical, rhetorical in cult recruitment, and of course moral. Natural philosophy yielded entirely to moral philosophy. Different levels of commentary encouraged a frame of mind that for about 1500 years found hidden meanings everywhere. That converted natural phenomena into something entirely different from their animal, vegetable, mineral nature.

Philo Judaeus (20 BCE-50 CE) was an Alexandrian, who in reading material he believed to have come from Moses created the first known version of combined theology, cosmology, and metaphysics. Extracting doctrine from narrative incidents was a laborious process. That green herbs occur before grasses in *Genesis* meant to Philo that ideas appear to the intellect before anything appears to the senses: "For why, when he [Moses] has previously mentioned 'the green herb of the field,' does he add also 'and all the grass,' as if grass were not green?... The truth is, that by the green herb of the field, he means that which is perceptible by the intellect only, the budding forth of the mind. But grass means that which is perceptible by the external senses" (27). That may read like a parody now, but it was intended seriously. It reduces biblical episodes to a handful of repeated abstractions. Did Moses have two wives, one hated and the other beloved? Then the first must be Pleasure, characterized by lasciviousness, and the second Virtue, characterized by chastity. Put lasciviousness at the head of one column and chastity over another and start sorting incidents. With just those two categories the reader can accommodate any number of them and link them to metaphysics as well as to ethics, the universe being divided into good and evil, things of the spirit and things of the body, light and dark. In Gnostic and Platonist fashion, material is bad and spirit good, a reversal similar to Plato's defining of the non existent as the truly existent and of everything that does actually exist as an illusion. Nothing in the universe is unrelated to humankind. When the first humans commit a sin in *Genesis,* worms in the ground and eagles in the air suffer for it from then until the end of time. When it comes to nations and cults, Hebrews and Canaanites become the exemplars. The goal is to eliminate not just lascivious pleasure but idolatrous and besotted adversaries. The eradication ordered by the war god is the nationalist and foreign policy part of the agenda.

Translations of events into doctrine is no simple task for someone with the entire Hebrew anthology to convert. The C. D. Yonge translation of Philo fills some 863 double column pages working out the details, the equivalent of a couple thousand normal pages. These are typical of confections that run through most of literate human history. Historical incidents continue to fall into place under similar exegetical principles throughout the allegorical tradition, determined as it was to make nature, historical events, legends, and people conform to pattern. When Paul and other New Testament commentators resumed that task, they inspired the church fathers whom J. P. Migne collected in the Greek and Latin patrologies. These total 378 volumes, 161 Greek and 217 Latin. Why Moses neglected to provide a key to his texts and Christ failed to explain the doctrine that interpreters extract from his teachings the interpreters aren't obliged to say. If the incidents themselves are somewhat dark, the commentary has to be more enlightened than the text. Even more contrived was Berossus, who proceeded from geography, chronicle incidents, and crops in Babylonia to Oannes with the body of a fish, human feet and a second head, and so on without transition to men with two wings and two to four faces (*Ancient Fragments*, 25-26). Whereas Berossus could sound reasonable until he plunged into the deep, scarcely a word of Zoroaster and the Gnostics refers to anything recognizable. The same reverse logic is at play: what is manifest and tangible is dark, and what is invisible is enlightenment.

The allegorical tradition putting diagnostic reason to the service of mythopoeic imagination lasted a surprisingly long time, well past Bacon's championing of induction. Partial revelations produced an impressive Christian/Platonist literature in the English later 16th century and earlier 17th century. In George Herbert a suppliant's dialogue turns the gap between a simulated biography and its meaning into a riddle left to the reader to interpret:

> I gave to Hope a watch of mine: but he
> An anchor gave to me.
> Then an old prayer-book I did present:
> And he an optic sent.
> With that I gave a vial full of tears:
> But he a few green ears.
> Ah Loiterer! I'll no more, no more I'll bring:
> I did expect a ring.

The two parties speak different languages because they live in different time zones. Like some sixteen centuries of preceding Christians, the speaker has expected a promise to be fulfilled. By analogy to human marriage, a ring would mean that the gap between parties has been bridged. A bond as old as the Yahweh/Abraham covenant would at last fulfill its promise. What would be enough time spent in limbo to reach that conclusion remains unspecified, which is why hope and faith occupy the interval. Under civil law such a situation would be Kafkaesque: everyone is to be judged but at an unspecified time. In the civil judiciary, ignorance of the law is no excuse. The laws are available. In divine justice, ignorance is inevitable. No judgments can be handed down as precedents because no one knows for certain how generations of predecessors have fared or will fare. What are actually on record are delays in the promised end.

Henry Vaughan, one of Herbert's more attentive readers, works with similar assumptions about hidden meanings that start readers toward a realm they can't reach on their own. Sundays open a portal upon it:

> Bright shadows of true Rest! Some shoots of blisse,
>   Heaven once a week;
> The next worlds gladnes prepossest in this;
>   A day to seek
> Eternity in time; the steps by which
> We Climb above all ages; Lamps that light
> Man through his heap of dark days; and the rich,
> And full redemption of the whole weeks flight…
>
> The milky way Chalkt out with Suns; A Clue
> That guides through erring hours; and in full story
> A taste of Heav'n on earth; the pledge, and Cue
> Of a full feast: And the Out Courts of glory. ("Son-dayes")

The poet presumes only to exploit the guises the Word has already taken, not invent new ones. Whereas the Jewish promised land, Canaan, was a mapped country with farms and cities, the new one is only something like that one, how much and in what way isn't determinable. As the title page of one of Vaughan's volumes, *Silex Scintillans*, indicates, tropes are struck from the poet's flinty heart not by anything local or timely and not by previous poets but by divine lightning. Sunday is the special day of the Son, and that pun Vaughan extends to stars that become illuminations of the Son on a path to heaven marked out by the Milky Way. Vaughan wouldn't have known of other versions of the image, but the sun as a symbol of enlightenment goes at least as far back as the Sumerian Ishkur. Indeed, as excavations in Mesopotamia, Persia, and Egypt started finding in the later 19th century, much Jewish, Christian, and Muslim practice in translating objects and events into signs and symbols had similar historical roots.

In the West's incomplete transition to empiricism and science we find the recurrence of that and other ancient beliefs alongside natural philosophy and science. Wordsworth's pantheism and Emerson's transcendentalism suppose a spirit/matter separation similar to Philo's, Vaughan's, Traherne's, and the Cambridge Platonists' except for spirit penetrating more deeply into nature. Like matter and like death, evil to Emerson is a "nonentity" (65). The oversoul stationed in an undifferentiated "ocean of light" (245) is alone real, a 19th century transcendentalist revival of Platonism. Thomas Carlyle (1986), whom Emerson sought out on a trip to England as a fellow spirit, believed that the hero "is he who lives in the inward sphere of things, in the True, Divine and Eternal." In daily life, that sphere remains buried "under the Temporary, Trivial" (236), which includes everything available to direct observation. Transfers of power from the visible universe to humans are a recurrent feature of illusions from the pre Socratics to modern day cults, whose numbers polls indicate still make up a majority of some populations. The details of the moral philosophy that prevailed for centuries were mostly for others, but a vague sense that everything in existence was intended as it is remains and somehow bears upon mankind. The wholesale removal of populations imagined in biblical texts but never carried out has dwindled to a few militant nations and terrorist cults.

Stirring the Mix

The question for myth criticism is just how thinking in terms of miracles and myths could work toward systematic understanding when they are as different as algorithm and ritual. On one level superstition lives easily enough with realism in that even the most credulous mind has to be realistic to function. In Philo's extracting of moral and metaphysical principles from narrative materials we've seen one answer as to how incongruous things can be combined. Two glaring defects mark the technique: the dogma presumed to be embedded in the episodes could have been stated outright and saved volumes of commentary. And very little natural or human history translates plausibly into what interpreters expect to find there. Even so, *imagining* and *deciphering* aren't completely opposite. Whether in mechanical or personified terms, thinking about the source of a storm is thinking about the source of a storm. The mind goes as effortlessly back and forth in modes of explanation as it does in switching television channels from cartoons to documentaries. Comprehensive accounts of natural history, however, aren't allowed to invent what doesn't exist or to dismiss what does.

The exegetical tradition wasn't of course the only one that had a method to convert myth into philosophy. Aristotle's preference for a coinciding reversal (*peripety*) and recognition (*anagnoresis*) that includes thought (*dianoia*) assumes an emerging rationality based not on allegory but on depicted events. It was based on what was assumed from the outset to be literature, not history. In retrospect from the end, themes have been at work all along in an authored text. Oedipus has always been the son of his wife. He just didn't know it. How does that differ from allegory's removal of material nature to reveal something intangible? In Philo the purpose is embedded in the universe and isn't owed exclusively to the author. Much of Graves' *White Goddess,* too, is taken up with correspondences between letters of the alphabet and such things as moon cycles, vegetation, and phases of the creation. In scriptural terms it is a designing super intelligence, not an author who associates Jacob with Babel, Hur, and Moriah. In Aristotle it is solely the author who connects the dots and brings meaning and plot to fruition at the same moment, which ideally combines recognition and reversal. A related aspect of myth/philosophy interaction is evident in riddles, prominent as early as Sumerian literary genres. They are built into narrative and drama but can also be presented separately. That methodical rationality figures in the deciphering again blurs the boundary between myth and philosophy.

The vagueness of upper tiers in transcendental substitutions for natural history, paraphrased in Carlyle's case from Fichte, includes a spirit concealed inside or behind objects as well as located in another realm, immanent as well as transcendent. As if foreseeing the next two millennia, the Greek atomists were skeptical about any such upper realm. Drawing mainly on them and on Epicurus, the Roman Lucretius in *De Rerum Natura* did what he could to dispose of such realms by having history unfold in stages not of enlightenment but of civilization. As an early archaeology, Lucretius' concept of evolutionary advances anticipates modern concepts of the bronze and iron ages. Ovid's decline from a golden to an iron age goes in the opposite direction. In Lucretian terms technological progress began with the foragers and barbarians who collected into bands and settlements. To repeat, however: as is clear from the retrenchments and survival of believed myths, when I say *from* myth *to* philosophy and science I don't mean that civilizations went through a time of fabulous storytelling and then became realistic. The displacing of myths by science and philosophy has often gone a step forward and a step back. Although the general movement overall has been from one to the other, advances and retreats had already been several by the time of Thucydides. Ernst Cassirer (1944) locates the origin of the sciences themselves in mythic texts. In Athens adding one way of thinking to another meant going from Homer and Hesiod to Plato and Aristotle with myths still prominent in Plato. Magic, astrology, and myth not only preceded science in that view but in Cassirer led to it: "It was a false and erroneous form of symbolic thought that first paved the way to a new and true symbolism, the symbolism of modern science" (49). Paleontologists and archaeologists go far enough back to see the tangle of myth and philosophy in depth, charting prolonged ages of stone, copper, and iron. In one typical framework, Thomas L. Thompson in *The Mythic Past: Biblical Archaeology and the Myth of Israel* (1999) begins with excavated sites dating 1.4 million years ago. Practical and aesthetic production came in proto civilized forms from about 50-40,000 to 10,000 years ago, the point at which agricultural and urban settlements accelerated advances in arts, crafts, and technology. One of the arguments against the Hobbes and Rousseau concepts of an overall deterioration is Hallpike's observation (2008) that "it has actually taken the whole course of history to reveal the full potential of human nature and human abilities" (28). Occupational specialization advanced that potential into operative *techne*, a product of applied science.

What Karl Jaspers called the Axial age–when conceptualization emerged more fully–wasn't strictly Greek but multinational. Work by recent scholars assembled by S. N. Eisenstadt (1986) examines precedents in Mesopotamia, Israel, India, and China. Another way of putting it is that with the exception of practical skills the looser analogies and associations of *mythopoesis* preceded detailed categorization. Carl Sagan puts it another way in *The Demon-Haunted World* (1996) in finding demon haunting dominant in the absence of philosophy and science. Histories of ideas usually date the latter not from Aristotle but from approximately Thales (6th to 5th centuries BCE) and the atomists Leucippus, Democritus, and Epicurus. Arabic studies in mathematics, music, medicine, and astronomy began early and remained prominent even through the dark ages, which were mainly European. The Persian Avicenna (c. 980-1037) and the Spaniard Averroës (1126-1198) thrived in that era of repressed learning.

Among others who had considered the symbolic value of myth by Cassirer's time was the German philologist Hermann

Usener, a presence in Cassirer's *Language and Myth* (1946), one of the classic studies of how, in psychological terms, myths came to be, what they continued to do, and what use the early scientists and philosophers made of them. To Usener as to Cassirer, myth creation was a dynamic process in which the mind, reacting with excitement, produced narratives that answered as much to the excitement as to phenomena themselves. Deities hurling lightning bolts are better for some purposes than treatises on electricity. Aeolus has a dynamic presence that isobars and the Coriolis force lack. People may have difficulty entertaining two such radically different ways of thinking at the same time, but nothing prevents them from entertaining one as *poesis* one moment and the other as science and philosophy the next. Dedicated belief collects around some mythic figures whom Usener calls momentary gods and daemons. These are notable, Cassirer observes, for their "immediacy, with emotions of fear or hope, terror or wish fulfillment." Myths begin, as Cassirer points out, as a "spark jumps somehow across and the tension finds release, as the subjective excitement becomes objectified" (33). It projects a psychological condition into objects and circumstances. Usener's *momentary* gods become "gods of activity."

Using such figures to explain phenomena and events is only one reason for their existence. Group coordination and stronger ruling authority are others. Purely rational explanations based on cause and effect lack those benefits. How divinities are used to legitimize the ruling elite in seven representative if quite different civilizations Bruce G. Trigger (2003) explains in detail missing in earlier myth studies. Rulers finds validation in rhetorical substitutions in the seven cultures he chooses, the Valley of Mexico, Classic Maya, the Inca kingdom, Yoruba-Benin in west Africa, Egypt, southern Mesopotamia, and Shang China. In Arthurian legend, the quest for the Holy Grail combines with chivalric virtues and kingship, in that case in a conflicting secular and religious rule that in Sir Thomas Malory becomes a legendary part of the English heritage. In Spenser it combines monarchy's chivalric virtues with a mythical universe climaxing in the New Jerusalem, with some conflict between the individual quest of the Red Cross knight and service to the queen. What myth puts in terms of dramatic narrative undergoes, Cassirer finds, "a progressive organization and ever more definite articulation" (36). That fits the twelve virtues Spenser assigns to different knights and the historical allegory that likens contemporaries to heroes and villains. No actual person combines the virtues, but they fill out the idealized realm that prefigures the ultimate one.

The vividness of devouring time in Chronos, of the wars of Titans and the giants, of the Olympians, Dionysus, Eros, Nyx (night), and Athena wasn't enough to save such figures from being translated into abstractions. A more methodical approach pulls them out of their original cultures, as Bacon in *The Wisdom of the Ancients* (1609) assumes they should be. Deciphering the ancients, he protests, isn't like entertaining oneself with a toy. Pan is universal nature with such and such progeny, biform because of upper and lower worlds. Orpheus sings in two ways, one in civil and moral philosophy, the other in natural philosophy. And so on through Proteus, Juno, Cupid, and others. The aim is to turn each figure back into what gave rise to it in an empirical or historical setting. Once myths have entered the mainstream, the ratio of fact to invention has little bearing on their popularity and durability, which may or may not draw on translations into applicable abstractions of Bacon's kind. According to Xenophon even the highly rational Socrates found nothing inconsistent between academic philosophy and deference paid to the mythic Athena, not a personified abstraction in that case but a living goddess. Bacon too found nothing inconsistent between studies in divinity and studies in natural history, though he no longer tried to allegorize phenomena in the exegetical way. That was Galileo's point as well in defending a heliocentric solar system. It could not after all conflict with scripture if God was behind both. If nature and scripture seem to conflict, the reading of one or the other has to be in error. Since in his mind no doubt existed about earth's orbit and spin, the church authorities must be wrong in their reading of what they said conflicted with it, which was the sun standing still for part of a day. That could be allegorized but not taken literally in a heliocentric solar system. The basic problem neither side was willing to confront: the text was mythic, not historical. Nothing of the kind ever happened nor apparently did any of the slaughter of Canaanites. Exempting a chosen belief however improbable from critical scrutiny has a long history. Even in semi-modern Locke (*Essay Concerning Human Understanding*, 2008), despite distrusting custom, adopts the beliefs of his times. As for others, they were victims of the past, of "Education, Custom, and the constant din of their Party," which "gives Sence to *Jargon*, Demonstration to Absurdities, and Consistency to Nonsense." Custom is the "foundation of the greatest ... of all the Errors in the World, or if it does not reach so far, it is at least the most dangerous," because it "hinders Men from seeing and examining" (2.33.18). Like anyone else, a philosopher can accept what his culture holds dear without giving it a second thought. Quite different ways of thinking coexist in one and the same person. Technology replaces lamps and candles more easily than new beliefs replace old ones. Charismatic figures can pin theses to cathedral doors and change institutional history, but complete changes in ways of thinking are rare. Bernard Lang's (1983) reading of the Israelite prophets finds that true in something as basic as the growth of monotheism, which to begin with was a response to a military and political emergency(54), not a decisive dropping of many gods for one.

At a professional level, science and philosophy in due course did leave much behind, as astronomy outgrew astrology and chemistry outgrew alchemy. Removing magic from herbs left medicinal plants to pharmaceutical companies minus the magic. As Aldo Leopold remarks (1949), "to love what *was* is a new thing under the sun" (119). So is manipulating how posterity will regard us, what the atmosphere we see today was doing last week in the South China sea, and what the climate of the last ice-age was compared to now. Something similar to Thucydides' advances over Hesiod's mythic cosmogony and Homeric epic occurred in places other than Greece, as in the movement from Beowulf to Bacon by way of the chronicles of Edward Hall and Raphael Holinshed, and later, in a sense, in Kant's and Hegel's passing a secularized version of formerly religious terminology on to Marx and Engels. From there it worked into Foucault, Lukács, and others who "materialize the dialectic." The transition from doctrine with its transcendental realm to ideology with its evidence selection went through a stage in which the absolutes of theology became those of secularized ideology in something of the way providence became the invisible hand in Adam Smith. Marx and Engels deny creation and eschatology, but Kolakowski (1978) is correct to say concerning Engels that he translated Calvinist *predestination* into *destined* economic forces (342) and that "Marx's faith in the 'end of prehistory' is not a scientist's theory but the exhortation of a prophet" (375).

Any decisive end is a contrived one in the context of 10^{100+} years of matter/energy exchanges on the way to entropy or for any other projected end such as the universe pulling back in a big crunch. More exacting philosophy and science sheer off the layers of uplift terminology, confess that the beginning is too remote to be tested, and project a course for stars and satellites into mega trillions of years. Needless to say, nothing that extended can be assigned a moral level or human-oriented narrative. That is the most comprehensive part of the transition from myth and moral philosophy to natural history's cause and effect sequence.

Reified Abstractions

What remained of Plato's essences we sometimes hear expired in the 19th century with the discovery of evolutionary *metabole* or change. In the context I'm assuming, essentialism is another hybrid myth/philosophical concept that assigns verbal inventions an immaterial Reality and a separate realm for them immune to change. Gerald M. Edelman (1992) holds that the theory of natural selection fatally damaged "Plato's idealist notion of essentialism—that there is a world of perfect essences of which the exemplars in the concrete world are merely flawed examples. Species are not essences or types; they are the result of selection from variation" (212). That is true, but the fatal damage registers only in selected circles. It is also true that scientific proof isn't conclusive, as numerous critics of empiricism and the scientific method have pointed out, including Karl Popper, Rudolf Carnap, Willard Quine, and Peter Godfrey-Smith in *Theory and Reality* (2003). John Gribbin in *13.8* (2015) has more confidence in the findings and the theory than Ian Stewart (2016), who acknowledges the possibility that "the universe is a lot older than we think" (230) and the shape anyone's guess.

To realists and naturalists, essentialism never did seem a valid way to account for the myriad shapes and species of the planet or for what could be seen of the heavens. Plato himself is inconsistent in that regard imagining how what experience encounters relates to Ideas. In visualizing regional land formations, *Critias* verges on an almost modern concept of natural history except for being embedded in general decline: "By comparison with the original territory, what is left now is, so to say, the skeleton of a body wasted by disease; the rich, soft soil has been carried off and only the bare framework of the district left. At the time we are speaking of these ravages had not begun. Our present mountains were high crests, what we now call the plains of Phelleus were covered with rich soil, and there was abundant timber on the mountains, of which traces may still be seen" (111: b-c, 1216). That may seem an unusual concession to geographic change, but it doesn't affect the notion of permanent forms becoming distorted in their incorporation into matter.

What Socrates says for the sake of argument and what he stands behind aren't necessarily the same. The dialogues came at a time when it was unwise to promote him too openly in Athenian democracy. David Fortunoff points out (1993) that like Galileo, Descartes, Spinoza, Bertold Brecht, and others, Plato recognized the danger the state posed to those who challenged it too shrewdly.

Made too cogent, criticism was less tolerable than when it came from less imposing sources. "The philosophical-literary form of dialogue . . . may well have met the requirements" of free expression (*Plato's Dialogues*, 64). Xenophon's biography of Socrates bears witness to that pressure to pay respect to shrines. Even so Plato has Socrates propose the realm of Ideas too prominently and forcefully to suppose that he considers it merely hypothetical. As a subset of the proposition that virtue isn't teachable in an uncertain world, *Phaedo* alludes to a realm free of change that *Meno* cites at length (82-84). Several assumptions follow from that, among them that mankind inherently wants to return to preexistence where Ideas remain uncontaminated. Socrates famously uses preexistence to explain why an uneducated boy is able to judge the validity of math equations. How could he do so unless something stayed with him from preexistence? Plato lets obvious challenges to that go unvoiced, namely that the boy doesn't remember anything else from before his birth and not even math without prompting from Socrates. The correctness of 2 + 2 = 4 could be self evident enough not to need preexistence to account for it. Other children become fluent in Arabic or Chinese and recognize grammatical correctness and error with an equivalent facility. Even if disembodied spirits were issued tongues and ears, those languages among many others presumably weren't spoken in pre-existence unless Tower of Babel conditions prevailed there too.

Cratylus likewise ventures into origins, this time as the source of correct names. These aren't attached to things arbitrarily and differently in each language but are necessarily right, like the correct answers to math problems. "A true proposition says that which is, and a false proposition says that which is not" starts off with an unpromising tautology (lies lie and truth is truthful). That names come built into things is more substantial and obviously more debatable: "Not every man is an artificer of names, but he only who looks to the name which each thing by nature has, and is able to express the true forms of things in letters and syllables" (390e, 429). *True forms* refers to the undistorted imprint of Ideas. Gifted people, perhaps again dimly remembering their source, intuit what objects should be called, which seems to be their Greek names. Those who doubt their capacity to do that should defer to those who don't. A correct namer must somehow regain contact with a template that was momentarily absent in the infant lacking language and never returns intact to a forgetful adult who finds a name missing. The horde of names a correct namer intuits must also include proper names of people, places, and things. If these aren't arbitrarily assigned, the repository that holds them must be enormous.

Again the dialogue avoids these obvious points along with the general principle that names are arbitrary conventions assigned differently in each language. In effect, the data are selected and arranged to demonstrate what can't possibly exist if all of natural and human history is admitted as evidence. It remains a crippling problem that nature produces millions of distinguishable species and innumerable objects and shapes. Even though names cite species rather than particular weeds and bushes, there wouldn't be that many source Ideas without a realm for them as confused as the real world. Socrates as devil's advocate doesn't allow the *Cratylus* notion of accurate naming to go unchallenged this time, but he questions it on the grounds not that words are assigned haphazardly but that matter in flux betrays whatever nomenclature is applied to it: "If that which knows and that which is known exist ever, and the beautiful and the good ... also exist, then I do not think that they can resemble a process or flux" (440c). What drops out of consideration are the laws of matter by which the combining and dissolving proceed and the practices of linguistics by which names are assigned.

The importance of Plato to our present concern is the deliberative dialogue form and its academy setting. Pedagogy based on rational analysis marks a turning point in the development of topic breakdown and step-by-step examination. That Plato draws on myths as well in the reifying of certain words and assigns them an existence free of nature makes him a bridging figure. He is the fountainhead of systematic academy philosophy and at the same time a repository of transcendental fictions whose place of residence is discourse rather an actual separate realm. The invention of a place immune to decay couldn't be more opposed to the naturalist's conviction that, as Burroughs puts it in "Straight Seeing and Thinking" in *Leaf and Tendril*, nature has "no 'hocus-pocus'," nothing on the borderland "between the known and the unknown, or that justifies the curious superstitions of the past" (*John Burroughs' America*, 52). Twilight may be more elusive to human eyesight than common day, for instance, but it is no less real. To the naturalist the problem isn't the unreality of objects but the limits of the senses and names matched against sizes, numbers, and incessant change. The senses, Burroughs finds, "are too dull and coarse to apprehend the subtle and incessant play of forces about us–the finer play and emanations of matter that go on all about us and through us. From a lighted candle or gas jet or glowing metal, shoot corpuscles or electrons, the basic constituents of matter ... are immersed in a sea of forces and potentialities of which we hardly dream (278).

Despite the deficiencies of essentialism as a way to account for what exists, critiques of it were comparatively sparse before the Renaissance. The reluctance to forego it was related to faith in a prime essence or unmoved mover. If essences don't really exist, the absolutes of the *omni* vocabulary are called into question along with them. The two go together in assuming a place quite different from anything in the visible universe. It is the *omni* terminology that is the problem, since an all-powerful force couldn't be constrained by the matter out of which it makes objects. We are back to the best of all possible worlds. If everything from atoms on up assembles and disassembles, why would a truly all powerful unmoved mover opt for things that decay rather than for their permanence? Plato's answer is the artisan nature of the Demiurge. He doesn't summon something ready made but like the pot maker shapes clay. The inherent nature of clay determines what is possible. The options open to the creator, as Leibniz maintains, didn't include anything better. The Andean

Altiplano kills off the vicuna and the guanaco because matter insists on raising mountains too high for creature endurance.

Francis Bacon

Myth's overlapping and contradicting science and philosophy works backward from the Greek 5th century as well as forward. As we've seen, because of the need for practicality in daily affairs induction and scientific method would never have been completely absent even when myths dominated concepts of origin and purpose. It is also true that Mesopotamian nations made progress in engineering, algebra, and the observation of celestial bodies as well as in animal domestication and crop raising. Greek science made advances in several fields besides Euclidean geometry and philosophy, as did Roman naturalism in Pliny and in engineering. Mathematics, logic, astronomy, and medicine were practiced disciplines in medieval cultural centers despite their being dominated by the church. It wasn't the beginning of more systematic investigation that came with the Renaissance, then, but a resurgence.

In making *The Great Instauration* more an insurrection than a revival, Bacon set aside several strands of philosophy that had converged in the Renaissance including Plato, Aristotle, and Arabic sources. Without explicitly addressing the problems that arise from the co-existence of divinity and natural history, he separated them, declared studies in divinity already well advanced, and concentrated on empirical observation. That set a precedent for dualism that allowed little contact between empiricism and matters of spirit. It allowed myth and science an uneasy coexistence in philosophy for the next two and a half centuries. If they were kept in separate rooms they couldn't quarrel as they had in Galileo's tiff with the church. In contrast to Plato's *Meno*, Bacon finds knowledge of particulars reliable enough if barely. The mind is a "mirror or glass capable of the image of the universal world, and joyful to receive the impression thereof" (*Major Works*, 123). It also craves order and finds it in greater uniformity than it is in truth and "is far from the nature of a clear and equal glass, wherein the beams of things should reflect... rather like an enchanted glass, full of superstition and imposture" (227). As the preface to *The Great Instauration* puts it, the universe "is framed like a labyrinth," riddled with ambiguities, "deceitful resemblances," and things "irregular in their lines and... knotted and entangled." He leaves it to later critics to point out that observation never covers all possible instances and so leaves induction without certainty except in its use as disproof.

Even when precepts are scrupulously checked against evidence the mind can never completely avoid confusion. That more than one theory of gravity remains in play is due to that as well as to the interplay of cohesive forces. Heat can break the strong force and convert hydrogen into helium, and gravity can affect heat by pulling great masses into spheres and increasing interior pressure, but in combination mass, energy, and heat haven't allowed anyone so far to formulate a grand unified theory of everything. Add to that the mind's insatiable fondness for lies. As "Of Truth" puts it, if vain opinions, flattering hopes, false valuations, and imaginations were removed, the minds of a number of men would become "poor shrunken things, full of melancholy and indisposition, and unpleasing to themselves?" (341). The best safeguard at the level of theory is to reduce findings to axiom or aphorism. "Except they should be ridiculous," aphorisms "cannot be made but of the pith and heart of sciences; for discourse of illustration is cut off, recitals of examples are cut off; discourse of connexion and order is shut off... so there remaineth nothing to fill the Aphorisms but some good quantity of observation" (*Major Works* 234). The key is the "good quantity," which brings high probability rather than certainty. Seeing black ravens in the thousands virtually assures us that outside of South African any we run across will be black. Meanwhile no one has ever seen a Demiurge imposing an ideal form on matter.

Bacon had no way to anticipate how much conventional wisdom the method he was advocating would eventually displace or gauge how easy it is to say 'produce axioms' and how hard it is to do so. Devotion to observation put the Royal Society, Hobbes, Locke, and Hume in a bind between empiricism and Anglicanism with its residue of a moral universe. Those who followed Bacon never explicitly doubted that natural laws could be overruled whether one called what did so Fortune, Destiny, Fate, Providence, God, or an invisible hand, but the occasions for intervention diminished with the extensions of natural law. Newtonian gravity had no need of outside supervision. Deism, a prominent theological position in the Enlightenment and of the church-state separation of the US Constitution, set aside the more abrupt intrusions of deity in Judaism, ineffective ones in repelling invasions. The project that Hobbes, Descartes, Locke, and Hume undertook wasn't put in terms of reconciling myth with natural history, but that is essentially what 17th and 18th century philosophers set out to do, taking the cosmos, its divine guidance, and its purpose for granted and working around the contradictions.

Quite apart from philosophy but not from science and technology was the applied science of the industrial revolution, not underway for well over a century after Bacon. Technology and social order by the mid 18th century took a course not directly tied to that of science and philosophy but related to them in working the materials as the materials dictated. In Engels' cryptic summary based on Morgan, the sequence in social orders, too, was set in terms of evolution proceeding through three basic stages, savagery with group marriage (a communal ideal), barbarism with paired marriage; and civilization with monogamy "supplemented by adultery and prostitution" (105). Marriage provided only a limited index of that social evolution, but as a sample it was reasonably representative. Nothing from the sky except the weather watched over nuptials or family continuity, the basic unit. Rousseau (1755, 1997) anticipates Engels in concluding more broadly concerning the social order and the mode of production that "the moment one man needed the help of another; as soon as it was found to be useful for one to have provisions for two, equality disappeared, property appeared, work became necessary, and the vast forests changed into smiling Fields that had to be watered with the sweat of men, and where slavery and misery were soon seen to sprout and grow together with the harvests" (167). At that stage, "metallurgy and agriculture were the two arts the invention of which brought about this great revolution. For the Poet it is gold and silver; but for the philosopher it is iron and wheat that civilized men, and ruined Mankind" (168). Substitute industry and agriculture for iron and wheat to get the fundamental change from hunting and gathering to agricultural and industrial production. Planting crops in rows led to machines that could do the planting. That mechanisms with their belts and sprockets are highly regular led to robotic jobs for those who saw to them. The mode of production was again the key to the progress of civilizations accompanied by applied science. The necessities of environment, biology, and botany were the keys to species change. What was the key to those necessities? In general terms thermodynamics, nucleosynthesis, gravity at work in the related areas of cosmology, astrophysics, and geophysics, the collaborating areas of the great leap forward of the 1920s. Not until then could the pieces be strung together chronologically and myths be reassigned to their sources in authors at particular times and places. Their explanatory function may have been secondary to their literary, didactic, and rhetorical value.

Chapter Six

Myth, Epic, and Oracular Sermon

Amplified Genres

The presence of the supernatural in myth, epic, and oracular sermon serves as an amplifier and has vital narrative as well as psychological and social functions. Merely name a character Zeus, Yahweh, or Venus and the narrative changes significantly from what it would be if it were limited to a normal cast of characters. For just how decisive the insertion of what Derrida calls a transcendental signified can be, take Milton's Story of Everything. Nothing in either *Paradise Lost* or its cosmos in what Milton took to be reality escapes the knowledge and power of God the Father. Every detail is governed by the Father and Son together, including indirectly even the scheming of Satan and the fallen host, who must be allowed to run free long enough to make the world the recognizable one. The forces responsible for confusion and suffering are eventually overturned by the extraction of a higher good from their escapades. That it *is* higher Milton uses to explain God's letting devils run loose, one key to justifying "the ways of God to men" (1.26). What is wrong in the history the reader is currently undergoing is due entirely to the first woman hearkening to the serpent and her uxorious husband hearkening to her. The rest of the animal and vegetable kingdoms and to some extent even the heavens are involved in the repercussions. Not only "all our woe" but every other woe except that of the fallen angels is included until the restoration that "one greater Man" (1.4) brings about.

This storied world history dominated several millennia under Judaic, Christian, and Muslim variants as it still does in fundamentalists of the three religions and more casually in many others. When Milton introduces the Father/Son dialogues readers learn ahead of time how Adam and Eve will fare and how Satan will be handled. They know the damage will be repaired and that Christians will end up better than anyone would have had the fallen host not tampered with the system. The poem is devoted to making that as believable as possible despite the implausibility of two humans bringing about such wreckage. The foresight and the unequivocal prohibition have the effect of 'prescriptive sanctity', to apply Francis Cornford's phrase. Anticipation and fulfillment confirm something foreseen from the outset in the command center. It isn't possible to install the restored paradise to begin with and skip the history because it is the latter than requires explanation.

In openly fictional texts, moral questions and character tend to be more shaded and ambiguous than in oracles and in credited myths where people are either blessed or cursed depending on whether or not they belong to a given cult. Achilles and Hector are complex characters. With his brutal slaying of Turnus, Aeneas presents a somewhat clouded and tarnished heritage to Augustus and the Roman ambitions contemporary with Virgil. That isn't the case with the Hammurabi Code, Hebrew oracular sermons, and the laws and ordinances of *Leviticus* and *Deuteronomy*, where acceptable behavior has less flexibility and the laws and ordinances have the highest possible authority.

In any age or culture one obvious benefit of high level connections is that the circle of believers–the cult and sometimes entire nations–occupies the moral high ground. If from the outside the founders of the belief look deluded, from inside they connect cosmic powers to the nation or to a self identified group within it. Why that is so and why beliefs become communal and binding to begin with is a question for brain studies like those of Thomas Gilovich (1991) and Michael Shermer (2011), but in general terms it is for psychological reasons that people are vulnerable to knowing "what isn't so," in Gilovich's phrase. A Hindu reading *Paradise Lost* has less difficulty taking it as fiction than some 17th Protestants did and fundamentalists still do. Non Puritans would have granted the story in outline and in many of its details, as Milton's anti-Puritan readers did from Hume, Hobbes, Dryden, and Locke onward. Some of them may have regretted Milton's 20 years of political activity and prose works that lent themselves to a bloody civil war, but they valued the poetry despite its maintaining a similar 'only God is king' stance against monarchical pretensions.

    Reconstructing possible effects on readers long after the fact is difficult, but what a text implies is reasonably clear where we know enough about the context. Epics and oracular sermons have named authors and hence are more available to *Sitz im leben* studies than mythology normally is. Myths too once had social settings, however. We may not know what they were, but an author's address to an audience can tell us a good deal. What people are assumed to be thinking guides admonitions addressed to them. In Mesopotamia, the Anunnaki were thought to be represented in the Tablet of Destinies, which granted extraordinary power to those who held it. In the heroic literature of Sumeria, the first of its kind to be recorded, deities weren't restricted to influencing natural phenomena. Investing personifications with human qualities made them more available for state affairs. Utu, the sun (later Shamash), had social relevance beyond anything related to the sun.

    Milton's parody of warfare in Heaven and in the Messiah of *Paradise Regained* dissociates divine power from monarchical regimes and from the military. The first epic style warrior is the father of lies.

His heroic bluster and the passivist hero's rejection of empire later are a turnaround for the poet himself, recently a defender of rebellion and regicide. The Messiah rejects not only the conventional battlefield epic as a way to establish a kingdom but the concentration of power in a ruling elite. Combining spiritual and civil powers is the devil's work. The renewal of a paradisal order, transferred off the planet, begins in the modesty of the savior, who returns unarmed and unobserved "Home to his Mother's house private" (*PR* 4.638- 639). By then he has enacted a paradigm for followers to let kings rule as they might, a major step toward separating a spiritual realm from history of any kind, natural or human. What counts is the individual soul's reception of the Holy Spirit.

Revelations? Or Literary Influences?

Once Julius Wellhausen (1878), following the lead of Wilhelm de Wette, established the "Judaising of the past" (223) in post exile authors, biblical authors in some scholarly circles were no longer considered to be divinely inspired. What he writes of *Chronicles*–that it owed its origin "to a general tendency of its period" (224)–could be said of other prophets as well, or for that matter of any literature written to guide an audience. The address of writers to their times is equally clear in Wellhausen's treatment of the already-encrusted earlier texts that *Chronicles* revised. The repetition of "rebellion, affliction, conversion, peace // rebellion, affliction, conversion, peace" (231) was set up less to answer to an actual past than to teach.

Several critics have followed Wellhausen's lead, among them Robert R. Wilson (1980) and Niels Peter Lemche (2008). The exodus story "long ago fell into discredit among biblical scholars. The exodus never happened" (82), Lemche concludes. That moves it into the fiction column with cross references to rhetoric. It had a rallying purpose. To the Deuteronomist's original readers the exodus story presented an actual heroic past perhaps to be emulated in comparable campaigns against surrounding countries. The prophets are militant but not necessarily on behalf of current regimes, indeed often critical of them. That conflict is exemplified in Samuel's correction of Saul's error in not obeying the command to leave no one alive at Agag. In Lemche's words again, "prophetic literature represents a message against extortion and social disorder. Israel has deserted its God and turned to foreign gods, the chronic complaint of the clerics and later of church militant among Christians. Israel has also neglected its obligation to care for the poor and destitute... [the prophets] are reformers and represent a special religious and sometimes militant group within their society, the 'Yahweh-alone Movement'" (86). Other critics as well have commented on the dual function of myths as *endorsements* of ancestral or contemporary ruling orders and as *critical judgments* of them. To the prophets no national solidarity is possible without conformity in doctrine, perhaps the reason Horace from outside thought Jews zealous (*Sermones* or *Satires* 1.4.139) and credulous (*Sermones* 1.5.96). The endorsement/critique functions appear side by side again in Milton, whose Puritan branch of Christianity revolted against pomp and circumstance.

Revelations presented problems in credibility long before Hobbes and Locke set out each in his way to distinguish fraudulent from authentic ones. The Hebrew stigmatizing of false prophecy to begin with was designed to eliminate borderland cases and practices of cultures based on augurs, sorcerers, charmers, wizards, and those who make inquiries of ghosts (primarily Wilson's list, 161). The strident tone of the prophets and their hostility toward neighboring states is based on that distinction. As the promised kingdom in Judaism, and in the New Testament the end of the world, were delayed, the problem of discredited prophecy *within* the sacred circle grew more acute. Peter's distinction between true and false prophets was based on the presence of the Holy Ghost in the former: "For the prophecy came not in old time by the will of man: but holy men of God spake as they were moved by the Holy Ghost" (*2 Peter* 1.21). That ignored the impossibility of knowing which if any inspiration comes from that source and left the problem precisely where it was. We find it still there in Hobbes and Locke, and it never entirely goes away with any oracular sermon requiring confidence in the one who delivers it. Where the credibility in the speaker comes not from demonstrable knowledge and trust but from a voice out of the clouds it has to be taken on faith.

What was at issue with Peter was the already troublesome delay in the fulfillment of messianic prophecies. He was baffled by that and issued a strange defense of what was beginning to look like another broken promise, one of many in the succession of invasions and other setbacks Israel had experienced: "one day is with the Lord as a thousand years" (3.8). To maintain that odd idea, he had to assume that an omniscient intelligence was unable to adjust to solar-system time. Moreover, when Jesus promised that the new heaven and earth would come before "this generation" should pass, that all heaven and earth "shall pass away" at that time (*Mark* 13.24-31), he was not being the least bit equivocal but measuring time in terms of human generations. That not only ruled out any discrepancy between human and divine ages but contradicted passages such as *Matth*ew 24.36 that discourage exactly timed predictions. That "day and hour knoweth no man." In *1 Thesselonian* 4.13-18 Paul, too, promises that those now alive will rise to the clouds without dying. They were addressed to listeners by a speaker with a recruitment purpose. Faith in evangelical missions is a way to ignore the evidence reason consults. Its denial of evidence is selective, however. It doesn't remove the need to be practical in daily affairs. In most versions of it someone with cancer is still advised to see an oncologist rather than a faith healer.

Contradictions in the matter of timing join with quite different versions of the second coming, like a thief in the night, yet noisily with a trumpet blast and a loud proclamation. To the new cult just moving into institutionalized doctrine and administrative practices, *this* prophet was as irreproachable as any ever had been, and he possessed the self-declared power to bring the final solution himself. None of the other prophets made such a claim. Peter is thus confronted with a many-faceted incompatibility between historical events and the power behind them, a branch of the age-old conflict between things as they are and a source infallibly in charge that somehow can't to be held responsible for them. If the aim of the prophecy and the other passages was to recruit followers and quiet doubts, the rhetoric becomes much clearer. It is designed to reassure a potential following that it has made the right choice and at the same time threaten those who haven't yet done so. The latter will be caught disastrously off guard. *Beware* is again the message.

The immediacy of the threat has another explanation that makes sense in terms of the expected Davidic warrior and the promised geographic kingdom. After the crucifixion, the *parousia* version of the Messiah's return took a middle course between the Jewish military savior and what will soon become an entirely otherworldly one. A geographic location can be taken or lost bit by bit. No such graduation comes at world's end or upon death. One is the narrative of a nation. The other is the conclusion of the creation-to-apocalypse story ending in an entirely different realm. The *eschaton* (the end) in the *parousia* version still includes an army but now an angel army that can overcome the Roman invader. The savior in that interim stage comes on a cloud, a symbol of commerce between earth and heaven and of mysterious appearances out of the unknown, yet with terrible thunderous power upon arrival. Depending on the prophet and the rhetorical application, armed force may or may not accompany a Messiah. In the millenarian segment of the 1640s Puritan rebellion, an army of saints performed the holy mission of liberation without angel assistance and was set to reign a thousand years. That amounts to still another compromise between Judaic and Christian myth. The Puritans were Old Testament in terms of political activism but New Testament in terms of an ultimate strife-free place. They were to use the thousand years to convert the Jews and other non Christians. Those who didn't convert were to them as Satan's crew to the good angels and neighboring idolaters to the ancient Hebrews.

Recruitment based on a decisive ending loses credibility with delay but only temporarily. Another prophet and another promise soon replace the discredited ones. The outcome and the form of the message remain the same along with the narrative. The rhetorician/audience relation is renewed as if nothing had happened. Collective fervor has meanwhile become part of the complex. No one decides to believe in the impending end of the world without like-minded company, and it gets to be like-minded through indoctrination. So discredited has the end of the world become that for the most part the end and transfiguration have been individualized. No one can prove or disprove a promise the fulfillment of which is out of sight. Return visits come only through seances. If time stretches out to billions of years, no problem. The key transformation has been individualized. Each mortal goes into end things separately. The results aren't subject to review unless a Captain Stormfield devises a spacecraft capable of reaching heaven and reports back to Mark Twain. What are the living conditions truly like when billions by the dozens collect there? A predictive answer lies only in the math. Thousands of generations mingling have to be quite different from New Jerusalems usually imagined.

Commenting on the discrepancy between legend and actual history and between prophecy and what actually happens isn't as common in criticism as one might expect. The reasons aren't altogether clear. The usual practice even among anthropologists is to treat concocted history about the same as verified history. They do recognize the difference and frequently mention it but don't build on the implications the distinction has for how social cohesion works and what the advisoryor admonitory purpose of the message was. Wilson is following that live-and-let-live procedure in reporting that in Mesopotamia all phenomena were considered related, which makes a fox appearing in a Babylonian plaza related to a catastrophe that comes soon after. To the trained archaeologist and to some textual critics, that non sequitur isn't anymore to be challenged or examined for its rhetorical effects than the "divine spark of true knowledge" in the prophets, "brought to consciousness by the spirit of God" (3).

Yet *inventing* a causal sequence has a motive that *reporting* one doesn't. As part of an indoctrination program, it serves a different social function. That is the essential difference between an educational retrieval of actual events of the past and indoctrination based on fabrication. Failing to apply the difference creeps into the history of ideas as well as into archaeology. Amos Funkenstein (1986), for instance, in tracking expertly through a labyrinth of theological, scientific, and philosophic thought in the Middle Ages and 17th century, refrains from applying current knowledge that would put the texts in a quite different light. Yet it was for good reason they were at odds with one another and often confused. Without physics and astronomy and without acknowledging the implications of the bare forked animal, they were adopting untenable premises from the outset. Vico and Hobbes, two of Funkenstein's more prominent examples, look more plausible than they would if natural history were placed alongside them. Spinosa, Leibniz, Hobbes, and Vico attempt to resolve puzzles without the key factors needed to do so. Because they had no notion of how the planet's topography and climate came about they couldn't include that factor among the sources of human conduct. Natural history is also largely missing from critiques of anthropomorphism and hermetic wisdom in the studies of realism and pragmatism.

Incorporated Literary Kinds

Where narratives are as extended as those of *The Aeneid* and *Paradise Lost,* inset moments are defined by their functions within them, subordinate to the whole, including rhetoric as well as imitation. Identifiable literary forms that normally stand on their own are re-evaluated in the context, as Milton in the epics takes possession of nearly every literary tradition he touches upon. Pastoral is rooted in Eden rather than in Theocritus. The marriage of Adam and Eve corrects epithalamia and the unrequited love portrayed in idyls and eclogues. The translation from a terrestrial to a celestial paradise revises the death-in-Arcadia theme. The book of creatures shrinks to the few playful animals who entertain the first parents and have the attractions children will eventually see in zoos. The eventual outcome of death in Arcadia is a new paradise that has no further need of classical pastoral except possibly something like the "other streams" of "Lycidas." The battle scenes of epic are taken over by Satan and made into devilish mockery until the Son relieves the good angels of their misery and without using a weapon dismisses the fallen angel army. The power he uses bears no resemblance to the militant forces of epic. Dialogue, too, is conceived anew from the beginning in the treacherous asides of Satan and his followers, the divine dialogues, the pre and post fall talk of Adam and Eve, and pedagogy in the different styles of Rachael and Michael.

All such amendments of classical traditions including hymns, laments, and soliloquies fall under the aegis of the Light and world-making Word. They are apportioned by rank and standing, the lowest form of address ultimately being the serpent hiss that replaces hymnal celebration when the fallen angel chorus welcomes Satan back to Hell. Where the incorporated kinds may not have had any very pronounced didactic functions on their own, in the story of man's first disobedience they fall into place along with everything else. Virtually every scene is part of a systematic cross referencing of levels in an elaborate comparison and contrast scheme. Creatures are expected to keep to their assigned places. Trying to change it without permission disrupts the scheme. One violation in the hierarchy is all it takes to bring reverberations throughout the lower ranks, which may be why in both the tradition and Milton's thinking recriminations for the human fall have spread through the plant and animal kingdoms.

Parody, ironic scorn, and heroic bluster come with the fallen as further distorted versions of hymnal celebration and contrite confession, the latter saving some people but not the fallen angels.

In the wordplay of Beelzebub and Satan may be a hint of Italian improvised comedy, mimic clowning movements behind a figure of status, but if so the Word has the final say in the matter of what is humorous as well as what is heroic. Repentance generates mainly elegy and lamentation. The unredeemable have moments of lamentation and regret, but they are lacking the cathartic satisfactions of true confession. The conventions of lyric are related to moral instructions and come as outbreaks like Eve's address to the Tree and the morning orison that she and Adam address to the creation and its deity. Excess devotion to the tree sets an idolatrous model. Poetry in an elevated style is one appropriate human parallel to hymnal celebrations in the high style of the angels. The as yet unfallen Adam and Eve hit instinctively on proper forms of address. What Raphael and Michael bring from outside Eden is exceptional in adding to one level knowledge of another, but then so are visitations from the Holy Spirit even now delivering the poem to the blind seer. Why that must be so I'll consider later in the belatedness of the poet.

Something similar in cross referencing is true of biblical texts as well but under various authorship and across centuries. The cross references come from links to former authors, as later prophets assume Moses to be represented in *Deuteronomy*. The connections aren't as tight as those of a single author constructing a comparison and contrast network, but later hymns, psalms, laments, wisdom passages, and chronicles expand on the initial covenant and the intervening tribulations of Israel. The disparity between the covenant and the situation generates much of the Hebrew anthology. The apex of the hierarchy is the same as that of Christianity and Islam in that in all three nothing exists that can't be connected to the *omni* terminology of Yahweh, God the Father, and Allah. That is the burden of the oracular sermons as well.

In *Deuteronomy* enemies who rise up before the Hebrews are to flee seven ways. Their bodies becoming "food for all birds of the air" and beasts of the earth: "I will send my fear before thee," Yahweh promises, "and will destroy all the people to whom thou shalt come...

And I will send hornets before thee, which shall drive out the Hivite, the Canaanite, and the Hittite" (*Exodus* 23.27,28). With "curses, confusion, and frustration," he will set upon Israelites too if they have dared swerve from righteousness (28.20). If those within the fold weaken they will be smitten with blindness and confusion of mind (28.25-28). Militancy usually comes in answer to what the orthodox define as idolatry. Isaiah promises with respect to Tyre that the very earth will wither under Yahweh's assault (24.6). The Lord will heap "terror, and the pit, and the snare" on enemies (24.17) before he turns on the Israelites themselves. Do not idolize alien gods, Isaiah warns his listeners, lest "the anger of the Lord your God be kindled against you, and he destroy you from off the face of the earth" (6.14-15).

In such passages Jehovah puts epic forces in focus as Milton's warrior Son does in finishing the war in Heaven. The Hebraic texts devote the listing device of the epic roll call less to the naming of tribes and heroes than to the satisfying details of enemy mortification, stricken with terror and the realization that they have chosen the wrong divine backing. Plot continuity and thematic coherence depend on unquestioning devotion and genealogy. To be included in the tradition, a belated author must establish a line back to the beginning of the covenant. Not all narrative episodes need emphasize that, and in fact quite a few don't, but the greater context extending back to the creation is always there once the texts have been collected. The Dead Sea scrolls came too late to be included but they too connect to authorized earlier texts.

Mesopotamian myth and heroic literature isn't connected in that manner. The dynasties and their divine support shift from era to era as do the nationalities of the authors. Some enrichment of individual texts does follow, however, from repeated deities. Subordinate genres, including incantations, petitions, rites of expiation, hymns, and laments, find suitable places in individual texts without reference to an inclusive context. The connections are a matter of individual authors alluding to predecessors. Genres of devotion are nearly as prominent in Sumerian, Akkadian, Egyptian,

Hittite, and Ugaritic literature as they are in the Hebrew anthology and they too serve national purposes. The teaching and enlistment side of the texts is less prominent than that of oracular sermons, but as the stelae commemorations of emperors testify, it is there. Any text of elevated style, like any festival or ritual, fosters communal spirit. Whether readers and participants realize it or not, they are being enrolled. It will be collectively that they build or raze cities.

Imperialism

Among warrior kings who claimed epic grandeur, Tiglath-pileser's memorial inscription, a mini-epic if that isn't a contradiction, asserts that the gods have granted him "power and strength" and commanded him "to extend the border of their land. They placed in my hands their mighty weapons, deluge in battle. I gained control over lands, mountains, cult centers, and princes who were hostile to Ashur" (Arnold and Beyer, 137-138). Deities are prominent again in his fabulous hunting exploits: "By command of the god Ninurta, who loves me, I killed on foot one hundred twenty lions with my wildly vigorous assault. In addition, eight hundred lions I felled from my *light* chariot" (Arnold and Beyer, 142). That Ashur and Ninurta have collaborated in state expansion is the burden of the prologue addressed to them and to Enlil, Sin, Shamash, Adad, and Ishtar (137).

The last mentioned, Ishtar, is the mistress of tumult as well as sex and war. Almost equally prominent in myth, epic, tragedy, and oracular sermon, tumult tests the strength of heroes. Though carnage isn't as common in Mesopotamian myth as it is in Homeric and Virgilian epic, few surviving texts dwell on it as much as the simulated histories of the stelae commemorations. Because the texts are carved in stone, they don't have the magnitude of *Gilgamesh* and later epics, but they make a beginning. Conquest is the means of expanding territory, and since people defend themselves it involves killing them. That isn't limited to the ancients. In Mark Twain's last extended speaking tour he notices the work of imperialism worldwide and roundly condemns it. Imperialism has always used stereotypes to justify domination. In the Hebrew anthology Yahweh is usually considered the instigator, not the kings or patriarchs themselves. The assault on Canaanite cities and killing of men, women, and children followed a pattern common for at least two millennia. If archaeological findings are correct, no actual invasion took place, and it is possible that earlier tales of conquest too were at best half true.

In the invention of a heroic past to equal or exceed those of rival powers in their vicinity, Hebrew scribes, possibly in the reign of Josiah, made their ancestry the equal of Near East dynasties. Virgil does much the same for Augustan Rome in elaborating on Homer's Aeneas, son of Anchises and the goddess Aphrodite. Using primarily the Hebrew example, though several others existed, Christian and Muslim warriors have reiterated the fervor of the past. Whereas campaigns in ancient Mesopotamia were regionally limited, modern communication and transportation make similar ones international threats. Much less dramatically, the ruckus *Ezra* records in the rebuilding of the temple and purging of foreign influences reduces most of the contention to lawsuits. The "people of the land" outside Jerusalem are the chief offenders. The collaborative rebuilding of the city, an unusually positive rallying of people in a hard time, draws strays back into the fold. During the exile of the Jerusalem intelligentsia, some who remained in Israel have drifted away and must be shaken up and reorganized. Within the precincts of Jerusalem, the collective psychology of cult devotion takes hold, and even on the perimeter outside the walls cult members agree to put away their foreign wives and children and return home. The only real resemblance to epic is the gathering of numbers. That is true again in the recovery effort of the families Nehemiah summons to repopulate Jerusalem.

The ultimate victory that oracular sermons project as a lasting kingdom and Milton endorses in the conclusions of several texts is absent from mythology and classical epic. In the Baal cycle of Canaan, a moment's peace provides at best a hint of what could be. In the words of the virgin goddess Anat she will remove war, foster peace, rain love down on the fields (Arnold and Beyer, 53). The response is resumed combat. The oracular sermon in contrast raises the stakes both in the scope of devastation inflicted on enemies and in the rewards for loyalty. In operating under a single high authority it requires a transformation of normal human psychology into a soul battle or psychomachia. After an initial concern with the former days of kings Uzziah, Jotham, Ahaz, and Hezekiah, Isaiah turns to the destruction of the Assyrians. In this case Yahweh needs no assistance in inflicting "wasting sickness" and destroying the enemy's forests and fruit of the land (10.16). The devastation of Babylon will follow. As he approaches "to make the earth a desolation and to destroy its sinners" (13.9), Damascus will cease to be. Egypt will look up in terror to find him riding on a swift cloud, possibly the source of the *parousia* phase of the New Testament's visionary end, which in Christian terms exceeds all previous destruction and punishment of non cult members. At that terrifying sight, idols will tremble and hearts melt. The divine avenger will turn brother against brother, bring widespread confusion, dry up the Nile, and starve the people. Tyre, the Leviathan, and Ephraim will be similarly wasted.

Lesser disturbances turn up in Norse, Celtic, and Germanic mythology, in the cruel deities of Central and South America, and in a milder form in the Chinese concept of the bronze age Ti, head of an afterlife that establishes a royal court beyond turbulence. In the latter case, the carnage that precedes bliss doesn't take place on the battlefield but in the court itself. For a courtier's leadership to continue in a higher realm, his servants must be killed to keep him company. The social electromagnetic bond in that case isn't mutual attraction but tyrannical power commanding loyalty. Herodotus' Scythians follow a similar policy. Sacrificial offerings, too, are limited carnage presented as a means to bring lasting peace on the assumption that the god or goddess in question likes blood offerings. Invading Spain is the way to the *pax Romana* in the Roman historians, as the conquest of Italy is the way to Augustan law and order in Virgil. Prophecy, revelation, and sacred ritual are phases of the rallying devices that put communal illusions into action. Even where ideology replaces doctrine, vestiges of holy causes remain as they did in the Nazi linking of Jews to Marxists in the Third Reich's purge, which in terms of actual numbers slaughtered exceeded any other on record up to Stalin. Plunder was again a factor however it was disguised. (Some of the stolen goods were still being recovered half a century later.)

It is catharsis rather than lasting peace through enemy removal that characterizes tragedy, a literary form in which the protagonist suffers more than his adversaries. If any benefits come of his death they are shrouded in mystery. Consider Milton's Old Testament example, *Samson Agonistes,* and what comes to Samson's countrymen. Where prophecy and sermon look to explicit enlightenment and sometimes peace after turmoil, tragedy cuts off the ending and brings not celebration but choral lamentation:

> *Chorus.* All is best, though we oft doubt,
> What th'unsearchable dispose
> Of highest wisdom brings about,
> And ever best found in the close,
> Oft he seems to hide his face, But unexpectedly returns
> And to his faithful Champion hath in place
> Bore witness gloriously; whence *Gaza* mourn
> And all that band them to resist
> His uncontrollable intent;
> His servants he with new acquist
> Of true experience from this great event With peace and consolation hath dismist,
> And calm of mind, all passion spent. (1745-1758)

Unlike oracular sermons certain of their ground, in tragedy no one knows why the unsearchable works in such excruciating ways. Fate, another common theme in tragedy, here takes the form of the Lord's "uncontrollable intent," and understanding comes best "in the close." The chorus is limited in point of view because the paradisal visions of "Lycidas" and the two epics are out of reach. The *pax Romana* has shifted its ruling territory off the planet and what would have been the victim nations are spared. The divinity projected into events, however, is much the same, possessed of a desire to dominate. Natural disasters become preliminary vengeance inflicted on the unblessed. Natural blessings become rewards for obedience and dedication. Those in favor aren't asked to administer any of the punishment themselves.

Mythic Heroes and Rituals That Work

Until the turn of the 19th-20th centuries, scholarship concerning early empires was handicapped by the lack of texts, some of which apparently influenced the Hebrew anthology judging by the similar psychology at work. Despite that shortage of evidence, Franz Delitzsch in the Bible/Babel movement and Hermann Gunkel's account of folklore and legend in *Genesis* (1903), as I noted earlier, acknowledged Assyrian literature and art to a degree not endorsed by most other critics and anthropologists at the turn of the century. That Gunkel found Hebrew literature brilliant and its amalgamation of legends "and their infilling with spirit of a higher religion" remarkable signals a shift in the evaluation of biblical texts. What counts becomes as much literary value as historical reference.

Gunkel personally believed *higher* to be justified and often rises to the oracular sermon in style himself, but we've no way to tell whether one text encouraged its readers to be more worthy on average than another. A household with a Jewish father, Egyptian wife, and well-behaved children might be commendable. Breaking it up, as Nehemiah and Ezra urge in the reconstruction of Jerusalem, might be bad for abandoned wives and children. Loyalty and devotion to a cause, prime virtues within a cause, become bad when they lead to ethnic cleansing. Following the divine example's removal of non cult members from the earth is a strange practice to call virtuous or spiritual, but precisely that has happened again and again from Mesopotamia, Egypt, and other ancient civilizations to the present.

Criticism of mythologies hasn't generally emphasized the social context as Delitzsch, Gunkel, and company were just learning to do. Even in their psychology they have been highly selective, concentrating on heroic archetypes and ignoring the sadism that group ancestor worship has often incorporated–in China and Japan, incidentally, as well as in the Near East and West imperialist movements. Persuasion has many social, political, and psychological uses, but as a narrative form of vague origin myth can also be divided into topics of the kind Northrop Frye and Jay Macpherson (1962, 2004) provide as well as into the kinds of rhetorical devices devoted to indoctrination that I've been emphasizing. The intimacy between myth and ritual also clouds the picture. Since modern myth criticism got underway in the later 19th century, anthropologists have often assumed that myth derived from ritual. The question of priority aside, the connection is logical. Ritual is incorporated into mythic texts, and myths are reflected in ritual. Both attract individual minds to a group mind coordinated by rhythmic movement and by recited phrases heavy in prosody elements.

One plausible sequence would be first the personification of natural phenomena, then the development of ritual and myth, and eventually institutionalized doctrines founded on both. That the personifications immediately increase their domain from thunder and lightning to human oracles and programmed worship, a sort of social algorithm, is a clear indication of the magnetic draw of groupthink. Whether any given prehistoric culture followed that sequence is impossible to say, since both myth and ritual had unrecorded pasts. Cuneiform inscription came several thousand years after the first settlements and was coincident with the first empires, but both myth and ritual had to have existed long before Sumerian and Akkadian scribes recorded them. *Gilgamesh, The Epic of Creation, Erra and Ishum, Atrahasis, Babyloniaka*, and the Baal cycle of Canaan are sophisticated literary forms that include ritualized passages. They aren't likely the doing of individual poets who sprang up on their own.

It appears that personifying natural phenomena led to the establishing of shrines and of rituals addressed to storm and sun gods and goddesses, and we've seen how their adopting of imperialist ambition inflated demigod emperors. The pretensions universally depend on substituting a fictional world for the real one, as do the sacrificial rites of central American gods. An unrecorded sun myth could have resulted relatively quickly in the Mesopotamian Shamash or the Egyptian Re, or it could have taken centuries. The only social effect natural philosophy need acknowledge is that whatever induces people to think alike promotes their coordination. If it also prompts them to think in predatory terms it promotes imperialistic conquest. Willing cooperation assists administrative efficiency. Administrative efficiency builds cities and raises armies. Since the efficacy of a ritual depends on belief and willingness to show it publicly that willingness too is unifying. Not to stand up and be counted is to activate shame psychology. Reason and objectivity have little chance of being heard under the circumstances.

In *The Hero: A Study in Tradition, Myth, and Drama* (1956), Lord Raglan (Segal, 1990) championed the anteriority of ritual, albeit acknowledging a shortage of proof. With Carl Jung, Mircea Eliade, Ernest Cassirer, and Joseph Campbell, psychological and archetypal studies and some philosophical ones complemented myth/ritual ones. Ritual has some of the stage-by-stage movement of mythic narrative but lacks the characters in conflict who stage the hero's removal of demonized rivals. Where antagonists appear they are defeated abstractly without the tension that accompanies epic upheaval. Mythic battles, too, are closer to ritual demonstration than the nationalist wars of the Deuteronomist or the carnage of Homer and Virgil. In imagining gods who stir up seas, raise floods and storms, destroy crops, burn forests, and inflict famines and plagues, myths assign them both cyclonic power and regularity. That is much as nature mixes springtime growth, crops, storms, and floods.

Inspiration and Belatedness

Virgil, Lucan, and Statius wouldn't have known about Homer's anonymous predecessors, but they acknowledged Homer himself. Their belatedness affects the stance they assume in announcing the themes of their texts. Virgil's *"Arma virumque cano"* ("I sing of arms and the man") responds in Latin to Homer's Greek. In the *Song of Roland*, in Dante, Malory, Spenser, and Milton, arriving on the scene late extended the heritage. Dante's making Statius an honorary Christian parallels his use of Virgil as a guide through the Inferno. That selection of sources may have been due to *The Aeneid* and *The Thebaid* framing ethical issues in terms of individual merit and to the exegetical practice of turning classics into anticipations of gospel. Even so, Statius seems an odd choice, since it is the fate of the polity rather than individual warriors to which the bird flocks of Melampus and Amphiaraus (3.500-548) refer. A reluctant city that takes no joy in warfare has no "fire or spirit" (4.345-355). Another reason for Dante's choice, however, may have been to underscore the limitations of classical authors. They can go only so far before someone possessed of revealed knowledge takes over and provides a transcendent level, the paradisal one that raises bliss from known levels to everlasting bliss. The angelic host is ranked around a mystic center in a state of peace and in visionary splendor beyond classical and Hebraic literature. The choral configuration is unlike anything in the epic battlefield hosts except in numbers and elevated style, again, however, with numbers only a fraction of what they should be unless Hell absorbs the surplus.

As an even more belated poet, Milton likewise displaces classical epic and versions of the heroic situated in monarchical and courtly settings in Malory, Spenser, and Shakespeare's English battlefield plays. Converting the traditional muse into the Holy Spirit allows him to detour around patronage and what he considered a compromised literary environment. Expecting to become a conduit of the sublime equal to biblical precedents isn't necessarily egomaniacal, since in *De Doctrina Christiana* inspiration is required merely to read scripture. No one can rise to a higher realm without an uplift that nothing historical or institutional can provide. In "At a Solemn Music," "Il Penseroso," and *Paradise Regained* the conclusion is cast in choral amplitude similar to that of "Lycidas," where those who dwell with God "weep no more," as in *Isaiah* (30.19). That departs less decisively from Hebraic prophecy than one might expect, because once the purge is completed in *Isaiah* not only will Egypt, Tyre, Assyria, and Babylon have ceased to be but the wolf will dwell with the lamb and the cow and bear feed together (11.6-9). Celestial paradises usually leave out the livestock and wild animals plus millions of other species of land, sea, and air.

The reception of the muse or spirit may pull the recipient abruptly out of his or her life. That can be traumatic if not fatal. The bard of *Paradise Lost* isn't subject to that in part because his blindness and the failure of the Puritan rebellion have already set him apart with something better than ordinary company:

> So much the rather thou Celestial Light
> Shine inward, and the mind through all her powers
> Irradiate, there plant eyes, all mist from thence
> Purge and disperse, that I may see and tell
> Of things invisible to mortal sight. (3.51-55)

Through all her powers is crucial. If the poet is to fill in where incidents are missing in scripture his mind must be cured of deficiencies and enlightened from an unequivocal source. Knowledge of divine dialogues, the creation, and Satan's counterplot would otherwise be as forbidden as fruit from the tree of knowledge. Where the deities of Homer and Virgil determine battles and national destinies, Milton's trinity is responsible for the entire universe. What is revealed to the poet includes events that happened before mankind existed and are not so much as hinted in scripture.

Even in the classical tradition where, except for Hesiod, revelations are less inclusive, some who are chosen as vessels don't take them on lightly. In the extravagant fifth book Lucan's *Civil War* (1992), a divine vision brings extreme trepidation despite its recipient having observed the appropriate ritual. Wanting to know his future and thinking to forward Fate, Lucan's Appius visits the caves of Phoebus and forces Phemonoe ("prophetic mind") to make herself receptive to Apollo–to be "received in virgin's breast... as the Sicilian peak gushes when Etna/ is pressured by the flames" (5.97-99). Knowing the danger of becoming such a vessel, Phemonoe refuses at first. When Appius insists, she is driven out of her mind and runs wild through the cave "dislodging with her bristling hair the headbands of the god/ and Phoebus' garlands." Only in madness can an ordinary mortal receive such an extraordinary vision:

> All time converges into
> a single heap and all the centuries oppress her unhappy breast,
> the chain of happenings so lengthy is revealed and all the future
> struggles to the light and the Fates grapple
> as they seek a voice; everything is there: the first day, the end
> of the world, the Ocean's size, the number of the sand. (5.178-182)

Any mind would have to strain beyond its limits to number the sands and foresee events to the end of the world. Some vessels responsible for visions of such magnitude are struck dead outright,

Mythic Worlds and the One You Can Believe In

"Because, if the god enters any breast, an early death is the penalty the human framework falls apart under frenzy's goad and surge, and the beatings of the gods shake their brittle lives" (5.116-120). Even as Lucan distinguishes between inspired godliness and madness, his epic benefits in scope and intensity from Phemonoe's fury, a variant of the tumult/mystery complex. Where the gods descend, amazing things happen, good and bad. Whatever else happens, the text gains in broadcast volume. That Milton went large without going mad was one reason that long poems for the next century turned mainly to satire. That's what was left before the romantics returned to nature as the scene of sublimity. We know from the rhetorical tradition that broadcast volume doesn't depend on an actual evangelical or enlightening vision. It comes with public address and is put to the service of any number of causes, to no more than the powers of television in the 1976 filming of Paddy Chayefsky's screenplay *Network*, which has numerous examples of harangue designed to raise not armed rebellion but anger expressed as shouted ritual slogans.

Delivery of enlightenment from beyond does nonetheless call more strongly than most causes for amplified voicing. It translates into rhetorical authority where the poet or the harangue endorses something specific. Milton's narrator calls out warnings as if to mankind in general. Lucan grows melodramatic. Turbulence is only to be expected where a newer order replaces an older one. If the message isn't convincingly delivered, the prophet will fail in his calling. History has many an unheeded prophet considered to be mad or fraudulent. Susceptible followers, not so much the prophet himself, make the difference. Some who were eventually accepted were initially doubted on the street, as Mohammed was in Mecca. Determining which annunciations and which texts are authentic has an all-or-nothing polarity about it. Not to be chosen is to be cast into outer darkness or taken as an anti-Christ, or what amounts to almost the same thing, being excluded from the canon and left unread. *Apocrypha* have nowhere near the readership of canonized literature. Because of their truth claims they were excluded from *poesis* from the outset. The lines between authentic prophet, madman, and sham are drawn in sand. Time that fails to bring prophesied events to pass erases them. If a prophecy comes with an imminent date it is sure to collapse.

As we'll see later, Hobbes and Locke worry that distinction between orthodoxy based on authentic revelation and shams like the Puritan revolutionaries. Allowed to live and write the epics after the restoration when many were imprisoned or executed, Milton presented a dilemma for conservative royalists. He did a good deal for the doctrine that they, too, endorsed and was a prominent follower of the epic and lyric traditions they honored, yet he had been among the 'enthusiasts' who claimed to be personally touched by the Holy Spirit. Among the greatest of English poets, he couldn't be discredited. Where the prose was sometimes seditious the poetry wasn't, and the Messiah of *Paradise Regained* had the saving grace to retreat altogether from politics. By limiting the feedback to an internalized spirit and not activating it in imperialism or a revolutionary reform, the poet could benefit from the epic expansion without presenting a danger to those in his immediate vicinity.

Disbelief

One needn't be too deferential about assuming that skeptics and realists lived side by side with ardent believers even where they left no texts. At least some of the time in their original settings as now, hyperbole would have been recognized for what it is and perhaps the stereotyping of other populations as well. Trigger (2003) cites newly conquered tribes among the Incas who remembered that before they were ruled by the sacred Inca king "their world had functioned perfectly well without him" (250). Going against the grain can carry a stigma as whistle blowing still does, but that is by no means universal: "If fear of divine sanctions alone were an effective control," Trigger continues, "why did all ancient states have to guard temples, palaces, and government storehouses from thieves and threaten robbers with drastic physical punishments?" (250). Good question. Persecutions come with holiness as book burning did with the rabid friar Girolamo Savonarola's reform in Florence. Bonfires lit by reformers and 'high places' closed under Josiah's command indicate that at least some of the population was considered heretical.

In reading biblical literature, critics like Robert Alter, Harold Bloom, Kenneth Burke, Herbert Schneidau, Meir Sternberg, Gerald Bruns, Frye, and Harold Fisch have taught us a good deal about poetics but typically less about the circumstances of composition, contemporary agendas, and the extent of dissent, which is admittedly hard to gauge. Archaeologists, historians, and some textual critics have done better in that respect. The militancy of oracular texts would have little purpose unless their authors considered disbelief and rival beliefs to be serious problems. Generally speaking, Hebrew prophets considered non believers abominations. The Muslim practice was to consider them infidels. Medieval Christian portrayed them as dragons, serpents, and other monsters, burned some at the stake, and stretched others on the rack. As Michelangelo's fantasy painting of Saint Anthony's torments illustrates, saints and martyrs were victimized by disbelievers. That is closer to myth than to epic, since the hero suffers passively.

Where codified doctrine, laws, and ordinances enforce orthodoxy, dissent is likely to escalate. It no doubt occurred to some under the Hammurabi Code to question whether the gods really wanted the hands cut off a child that struck its father. Or wanted the daughter of a man who had fatally struck an upper class women to be executed in the man's place (Arnold and Beyer, 113). Dissent can crop up anywhere, within an ecclesiastical order, in the palace revolutions of courts, and within families. From the beginning of Israelite kings, Saul and Samuel as representatives of civil and religious orders were at odds. Divine commandments allowed no margin of error: "Now go and smite Amalek and utterly destroy all that they have, and spare them not; but slay both man and woman, infant and suckling, ox and sheep, camel and ass" (15.3). In the opinion of Samuel, Saul is soft on idolatry. Samuel's berating of him (*1 Samuel*) for sparing King Agag is clear about what must be done to reconnect the current kingship with the legacy of Abraham and Moses. As a priest he carries out the duty Saul has shirked: "And Samuel hewed Agag in pieces before the Lord in Gilgal" (15.33). *Before the Lord* is the key: to be sanctified, any act must conform to divine mandates. These are delivered to prophets and carried out by priests like Samuel, whose fury thereby becomes holy fury.

Whatever the level of disbelief among contemporaries, it is greater outside the culture where little that claims to be inspired looks genuine and much looks spurious. That goes as much for those who write of Enlil, Marduk, and Baal as it does for Lucan writing of Phemonoe and Apollo. Belief eventually becomes a non factor as a poem ages and it is clear that it is Homer who assigns plot-forwarding roles to Poseidon, siding with the Greeks, and Zeus siding with the Trojans. He it is alone who assigns strategy to Zeus and Hera and field operations to Poseidon and Apollo. At the outset of *The Iliad* the narrator announces his subject in what became a conventional announcement of epic subjects of magnitude, again with no durable belief factor and perhaps not much even in contemporary audiences:

> Sing, goddess, the anger of Peleus' son Achilleus
> and its devastation, which puts pains thousand fold upon the Achaians,
> hurled in their multitudes to the house of Hades strong souls
> of heroes, but gave their bodies to be the delicate feasting
> of dogs, of all birds, and the will of Zeus was accomplished
> since that time when first there stood in division of conflict
> Atreus' son the lord of men and brilliant Achilleus. (1.1-7)

Putting gods in the picture raises havoc as well as voicing to a high level, precisely Milton's point in presenting the terrible aspect of the Word casting out the fallen angels and tormenting them as hissing snakes when they think to celebrate Satan's success. Putting false gods in control leads to the imperialism he has Jesus reject in *Paradise Regained*. In the death of heroes, tragedy and epic would converge but for epic's panoramic scope. Without the horrors, the combat would be less intense and heroes less heroic, as tragic plots would be less cathartic if pity and fear were diluted. The psychology of sacrificial offerings calls for that expenditure of emotion on the way to "calm of mind, all passion spent."

A Suitable Antagonist

If something is holy, something else must be unholy. That is a ground rule of literary texts and some psychology, and that too increases the furor. When it came into effect is impossible to say but a good guess would be with semantics, where definitions include, exclude, and polarize. Synonyms and antonyms may not have arisen from observation, but once they were established they could be combined with fictions to sharpen likeness and difference, protagonist and antagonist. Epic needs heroes, and heroes need opponents. It is part of the neatness of language to make clean categories whether or not history provides comparably decisive boundaries. The difference between a Parisian and a New Yorker may be negligible but the titles are insistent. Least filled with what might be taken for realism are ogre tales and parables, the episodes of which can be vivid without much detail and can polarize good and evil quite easily. In the manner of gargoyles and caricatures, belittling tales are sometimes directed at specific groups or classes, as in the portrayal of owners and overseers from the standpoint of slaves and laborers or of the morally depraved as seen by the righteous. Collecting ogre-like deformity into a single archetype produces a generalized scapegoat that can be made responsible for all the world's failings. Scapegoats need have no particular assigned character or single origin. The child-killing monster-goddess Lilith is quite different from misshapen ogres, more like the deadly beautiful witches of fairy tales. In the context of supreme fictions, the more powerful the deity, the more comprehensive the trouble maker has to be, unless, as in Malthus, tribulations are inflicted by divinity itself to strengthen character. Otherwise if saints exist so probably will devils and witches. During early Christian recruitment when the synoptic gospels were being written, the apostles demonized rival cults including Judaism and admonished converts to abandon what from then on must be regarded as heresy.

Elaine Pagels in *The Origin of Satan* (1996) charts the best known of the personified evils, Satan (al-Chadian in Arabic and Islamic tradition), following him through several incarnations ranging from an early Jewish superhuman to the author of all evil in Dante and Milton. In *Numbers* and *Job*, Satan is the virtual agent of the Lord himself, in other versions both an adversary and a collaborator. As we've seen in Milton, at the precise moment God makes the Son the highest communicant and the soon-to-be voice that brings the cosmos out of chaos, Satan springs into being as confusion personified. The negative element the Word lacks he must provide if the linguistic spectrum is to be filled out. Only when he is ushered in can language include sarcasm, scorn, baiting, name calling, strutting, posturing, feigning, hoaxes, fraud, falsehood, tall tales, simulation, mendacity, skullduggery, pretense, feigning, diddling (even taradiddling), swindling, rooking, toying with, forging, foxing, and fobbing. Multiplying that English vocabulary by thousands of equivalents in other languages gives us some idea how versatile *Homo ludens-diviosus-insidiosus-homicidus-sapiens* has been for possibly some 2000 generations since semantics got underway with its differences and likenesses. Satan or some mythic creature like him had to be invented, and along with him Mammon, Mephistopheles, Beelzebub, Behemoth, Moloch, and numerous others working behind the scenes. The cannon they invent not only parodies epic but puts communication in Belial's punning "terms of weight." The cannon's mouth opens and chaos rather than a proclamation belches out. Instead of dancing out choral responses, the good angels tumble in disarray. Such passages make *Paradise Lost* almost as much about language dynamics as about obedience and disobedience. The angels come equipped with just what they need by way of communication. Adam and Eve immediately hit upon sounds assigned to things, identified and named for the first time without a long apprenticeship in proto language.

Polytheistic monsters serve similar purposes in Greek myth, but the furies and harpies among them tend to be regional and aren't associated with national and categorical enemies. They are more avoidable than the universal Satan, though an angry Minerva or a Juno (in *The Iliad* and *The Aeneid* respectively) is far ranging. If Troy is to burn or if Turnus is to wage a senseless war against the invading Trojans, the reason is to be sought in bickering immortals, who, unlike figures in a psychomachian battle of spirits, don't come decisively good or evil. It is unclear finally whether they or Achilles and Aeneas determine the outcome, but from the viewpoint of literary criticism it doesn't matter: both serve as story-extending complications and create occasions for the display of courage, loyalty, and betrayal. Like Milton's anti-Word, such figures project the negatives of language that from the standpoint of linguistic history would have been inherent in semantics almost from the outset. True and false come as entwined as *accurate* and *inaccurate*, *integrate* and *disintegrate*, *construct* and *deconstruct*, *good* and *evil*. The words themselves remain detached from things, and their assignments are arbitrary. *Ceci n'est pas une pipe*, the Magritte painting of a pipe declares. Neither is the word *pipe* a pipe. If the interpreters are wise they won't forget the distinction between imagination and reality and won't confuse the need for language to fabricate words and relations with reality. The objects and relations exist as nature produced them. The overlay of language and math is meant to be as responsible to them as observation can make them. One of the main functions of critical reason as it is of literary criticism and theory is to check the match between the artifact and what is out there and where they diverge look for the reasons and extent. Persuasion and its rhetorical tools are likely to be prominent among the reasons. Literary and rhetorical devices hold hands when they bow to the audience at curtain time.

Repair? Retrofit? Or Replace?

Chapter Seven

Modernism and Beyond

The Four Century Renovation

Modernism and postmodernism look different if you come from a distance, from history and prehistory, rather than from in the midst, more lustrous, amazing in achievements, and revolutionary in outlook. They have also been spotted with atrocities enabled by harnessed energy and triggered by devotion to dogma or to doctrine incompatible with reality. That's if reality is taken to include such things as the periodic table, nucleosynthesis, and biological evolution, among the results of some four centuries of concentrated empirical investigation. Have these impressive achievements done anything to clear away persisting illusions? Yes and no. Decidedly yes for those aware of what the visible universe has been found to contain, mostly no for general populations.

For the first group, modernism's advancements in learning established a different address for all segments of the universe. In James Joyce's *The Portrait of the Artist As a Young Man* the home address of Stephen Dedalus goes from county to country to Europe and on out to the universe, which remains largely undefined. The layers now go from street to county to country to world, and out through the solar system and Milky Way to what in 1953 became the supercluster Virgo, about 100 million light years across. The fuller cosmic address now adds a supercluster containing globular clusters in an overall configuration that has been variously described as "meatballs within meatballs" and having the variable pocket density and filaments of a sponge (Gott, 56). Members of a race put in a less prominent place by that realization of the greater one are themselves like microbes in that great sponge. Passing through one of the galaxy clusters at nearly the speed of light would take several times longer than humankind has existed. Adding primate and quadruped ancestry still wouldn't get anyone through an entire galaxy supercluster. At that speed, stock a voyage with provisions for a couple billion light years to get a fraction of the way through. Overall the universe is now sometimes estimated to be over 90 billion light years across, the approximate span a spectator could conceivably take in by looking in both directions and adding the extra distance the first galaxies have traveled since their light started our way. Nowhere near that much is actually visible, of course, only something over 13 billion light years in any direction. The structure of what can be seen has been described variously not only as meatballs and a giant sponge but as a honeycomb and a Swiss cheese. The filaments in any case stretch in what look like great walls, one of which, computed by the Sloan Digital Sky Survey (the Sloan wall) extends 1.37 billion light years, not an easy adjustment for someone in James Joyce's time used to horse and buggy travel and not much easier in the age of jet travel. The difference between 3 and 3000 miles per hour is negligible by the standard set by the speed of light. To those in some areas of research the relocation of the planet is a major part of modernism, probably the part that has left the serious believers in cultural myths furthest behind. The representatives the latter elect are usually nurtured on the same archaic points of view they themselves hold. Like attracts like.

With rare exceptions like Giordano Bruno, nothing proposed in intellectual history resembled such a universe. (Bruno's ventures into the Copernican universe got him tried for heresy by the Inquisition and burned at the stake in 1600.) When misconceptions have been taken as truth long enough and made the basis of social orders, they generate jobs and positions of honor. In ecclesiastical institutions they work down into the details of parishes and congregations and into minds where they construct their alternative universes.

The term *modernism* itself had become so often used and debated by about mid 20th century that it spawned *postmodernism*, which applied as much to humanities and the arts as to the sciences. Being linguistic based, it added a distrust of discourse to the rest. Dragged through the tangles of Heidegger and Wittgenstein, language concerned not so much outside referents as other language and the problem of matching words to things when the things themselves were elusive. That was in defiance of the simpler S-V-O conjunction that connects real subject with real objects. In "I grab banana" all three parts of speech have tangible referents, unless they are being used as examples of grammatical conventions like this one.

After mid century randomness claimed an area of postmodernism as its own, and whether connected to that or not, chaos theory was born. To represent irregular non Euclidean shapes it put them in diagrams, Mandelbrot sets, language, and math had some success in diagraming a wider range of structures than traditional geometry had managed. No one could do anything with the completely detached particles of a dust cloud beyond approximating amounts and borders. The history of transmitted information from drums that talk to modernism James Gleick squeezes into a single volume in *The Information* (2011) albeit 526 pages. From the standpoint of science, information and discourse are the ultimate payout of organized atoms. Configurations in space come together under laws of attraction, trajectory, and speed but not under a schematic design that suggests a purpose. Two celestial bodies bumping together whether galaxies or dust particles are merely doing what they must.

By the later 1930s concepts of curved space, relativity, and quantum mechanics had filtered into some representations aimed at the general public, entertainingly starting in 1938 by the Russia born scientist Georgiy Antonovich Gamov, or George Gamow. That popularization, too, however, carried only so far. Plate tectonics, for instance, though the theory thereof was underway as continental drift by 1910, didn't reach public awareness in the US and elsewhere until the late 1950s. Even then, despite random land movement, that had little effect on prevailing concepts of intelligent design, a residue of the myths of origin. Neither, strangely, did the big bang have much influence on popular beliefs even when it was followed by the discovery of black holes, dark matter, dark energy, and gigantic areas of celestial debris.

The gap between those who realized the significance of the new discoveries and the rest had already begun to widen in the early 17th century, which saw the first serious reorientation of philosophy and science in a few troubled theologians, philosophers, and poets. John Donne sounded an early alarm over irregularities seen by sky scanners working so far without telescopes:

> The Sun is lost, and th'earth, and no man's wit
> Can well direct him, where to looke for it.
> And freely men confesse, that this world's spent,
> When in the Planets, and the Firmament
> They seeke so many new; they see that this
> Is crumbled out againe to'his Atomies,
> 'Tis all in pieces, all cohaerence gone;
> All just supply, and all Relation. ("The First Anniversary," 205-214)

Just supply would include what each needs compared to what nature provides, and *relation* everything coherence should connect. The alarm and skepticism of the crumbled heap were partly what gave Donne a modernist look in the 20th century rediscovery of the metaphysical poets. The disturbances and oddities that he laments suggested to some in the 17th and 18th centuries that a return to the *deus absconditus* or absent deity might work better to explain how nature had come to seem less provident. As the Dean of St. Pauls and writer of oracular sermons himself, Donne didn't take the usual overseeing deity out of the picture. Whatever the initial divine intent might have been, the universe must have fallen away from it.

For the next century after Donne, philosophers found themselves squeezed between the inductive method and what had derived from Pythagoras, Plato, Aristotle, Philo, sacred scripture, and philosophic theologians like Bernard of Clairvaux and Thomas Aquinas. One alternative for those who held to the status quo was to encourage distrust of science overall, accomplished brilliantly in Swift's Lagado episode. Science denial has sometimes made strange bedfellows, the strangest perhaps on the modern scene being religion and big finance. Where a specific matter such as the damage smoking inflicts on the heart and lungs is incontestable, calling methodical observation in general into question poisons the well. The data must be what is in error, not the makers of the product. Creationists entertain a similar broad distrust over a century and a half after Darwin in contesting the changing forms of life. The science denial in that case has to extend to astronomy, astrophysics, geophysics, geology, and DNA research.

Hobbes, Hume, Locke, Swift, Descartes, Leibniz, and Spinoza, each in his own way, entertained a more limited distrust of sensory impressions while trusting in the creation despite the irregularities of biology, prey/predator cruelty, punishing climate, and rough terrain. That remained true of commentary on nature through much of the 19th century despite what Malthus and Darwin pointed out about the severity of population thinning. Modernism on several fronts made it increasingly hard to hang on to any version of that stable and generally well made universe. How philosophers tried to combine old and new makes a curious chapter in the history of ideas. They are instructive for that reason and for what they illustrate of the persisting reluctance to accept natural history for what it is. In what followed

Bacon's proposal for the advancement of learning, philosophy became a patchwork of empiricism and spectral spiritual presences. Angels still hovered in portraits. Castles and country mansions were maintained as if a great chain of being were still the presiding order.

In a remarkably short time, less than a decade of the 1920s, the turn about was basically completed. The history of science and philosophy looked quite different from the vantage point of their new cosmos and its physical makeup from particles to receding galaxies.

Thomas Hobbes (1588-1679)

Hobbes and the others I mentioned had no way to anticipate species put in motion or what the 20th century consortium of astronomy, geology, physics, chemistry, and anthropology would find in their respective areas. Hobbes illustrates as well as anyone the difficulty of coming to terms with what in the 17th century was already a revolution on several fronts–global exploration, a solar system at odds with the geocentric tradition, and social upheaval in the rise of rebellious Puritans and commercial interests. The empiricist branch of philosophy assumed for some time that studies in natural history in the long run would agree with the providential universe of both Puritans and Anglicans. As some Royal Society treatises and Thomas Sprat's history of the Society revealed, those who were most aware of flora, fauna, microbes, and geography faced similar difficulties in reconciling received dogma with nature. The attempt was nonetheless to do so. That not everything could be salvaged logically or to the satisfaction of church authorities was what burned Giordano Bruno and landed Galileo in house arrest until his death in 1642. Explaining sea-life fossils in rock at elevation required ingenuity under any currently available earth history, though the biblical flood helped in that case.

Thanks partly to the unreliability of the senses, one of Hobbes' key assumptions, the study of natural history was prone to error, though none of the important findings of continental scholars and the Royal Society could be attributed to that. Neither of course could explorations at sea and on shore around the planet, which included frozen and burning wastelands as well as craggy topography. The most prominent and unavoidable evidence of natural roughness was the wrinkled, eroded surface of the planet itself, but so dominant was the habit of placing of mankind at the center of things that imperfections of that kind and disease among people and animals continued to be charged to the initial act of disobedience in Eden. A reliable natural history case wasn't possible in any case without atoms and thermodynamics. Physics, geophysics, and astrophysics wouldn't join forces for over another two and a half centuries.

One difference in Hobbes lay in his calculated degrees of sensory unreliability in which errors become more prominent as impressions weakened over time. Imagination and memory stem from decaying sense, and dreams are "the imagination of them that sleep" (5). That dimming of impressions results in imagination's "castles in the air, chimeras, and other monsters." That limited psychology, however, left out the main engine of the distortions, the self centered orientation of the ego. What Hobbes considered monstrous included beliefs recorded in ancient texts and the dogma behind the Puritan rebellion. That people are vulnerable to these Hobbes makes responsible for seditious factions stirred up by false prophets. I won't go into consciousness working on impressions in that psychology, but for several reasons variations in impressions are less damaging than he assumes. Color from flowers, sounds from a bell, and reflections from water are secondary only in function. The color of hair isn't as essential to bodily functions as the heart and lungs, but then neither is a toe lost here or a finger there. That doesn't affect the reality of what is functionally expendable. Being prominent or incidental has to do with taxonomy, not with what is and isn't real. As the biologist Ernst Mayr (1964) comments concerning systematics, "The most practical diagnostic characters are those that relate to some easily visible structure and low variability" (19). What may be of no particular functional importance can serve as a species identification marker. Something comparatively minor at the moment might be the key to future survival or species modification.

No one for some time would be in a position to begin at the indivisible base that Greek atomists considered essential to the diagnostic process. Like Robert Boyle (1627-1691) and John Locke (1632-1704), Hobbes begins with already compounded objects in which structure looks to be more substantial than surface elements. At the atomic level, however, reality doesn't come in degrees except in the sense that an object is more real than simulacra made to resemble it. Mockups and models occupy just as much space and may be just as substantial as the real thing. In the postmodernist concept of information, that's what they are as molecular composites illustrating of replication and organized systems. At bottom the distinction between primary and secondary features may be merely statistical. Horses' heads are greater in number than horses with black hair. *Hard, soft, tall, short, fast, black* are not expendable, merely variable. At what point does a mouthful of hay become part of the horse? Nature doesn't really care except in dispatching nutrition along vascular paths. At what stage did apes become bipedal hominids? Nature doesn't care about that either, only that each stage had to be capable of reproduction and adaptable to its environment.

Incessant change may puzzle philosophy unversed in DNA, neutrons, and electrons, but what it calls for are laws of change and an account of incremental stages, not different levels of reality or a master plan. Diagnostics may seek out what is functionally vital, but systematics looks for the relation of all parts however minor. The total of their relations amounts to a complex web or labyrinth, and the relations may be point by point or slightly overlapping, not systematic. The filaments of galaxy clusters look to be arranged overall in that way, though of course operating under invariable laws of physics. They simply have no overall cohesive force placing them in order. No method of investigation can map such a web except in a general way, deleting most of the detail. Biology has no need to go below the cellular level into molecules except in tracking DNA instructions for reproduction, as road construction can ignore earth's molten layer. Anything designed and built by humans has a purpose that governs the information it gives off. Anything structured by natural laws has no such purpose. Its information consists of the structural relation of parts where it exists.

The variable strength of sensations aside, Hobbes found errors as inevitable as mistakes in addition and subtraction. Thanks to the ease with which words put misfits together, no other creature is as vulnerable to error as the symbol user: "Metaphors, and senseless and ambiguous words, are like *ignes fatui*; and reasoning upon them, is wandering amongst innumerable absurdities; and their end, contention, and sedition, or contempt" (28). Again the leap from a fleeting sensory misapprehension to sedition as if mistaking a mirage for an oasis led straight to regicide. The sedition part of that clotted statement is emphatic in *Leviathan*, a treatise as much about mob behavior and the Puritan rebellion as it is about psychology and things viewed empirically. Vulnerability to lies and to what Hobbes calls uncertain science, meaning, at the extremes, vulnerability to absurdity and madness, stems from words broken loose from sense. The recent regicide has made what was *devilish* out to be *godly*. Nothing can be godly that results in deposing a monarch. Imposters were calling themselves saints, which is typical of insights into the myths of others and blindness to one's own. In those whom Hobbes called the gentiles, illusions had always been richly and variously nonsensical: "There is almost nothing that has a name, that has not been esteemed amongst the Gentiles, in one place or another, a God, or Divell; or by their Poets feigned to be inanimated The unformed matter of the World, was a God, by the name of *Chaos*. The Heaven, the Ocean, the Planets, the fire, the Earth, the Winds, were so many Gods. Men, Women, a bird, a Crocodile, a Calf, a Dogge, a Snake, an Onion, a Leeke, Deified" (74). With respect to pagan myths, Hobbes is content to assign them to poets, slightly mad poets. The exception is not surprisingly Judeo-Christian doctrine. The six day creation,

Jonah swallowed by a whale, the flood myth, the parting of the Red Sea, and the virgin birth belong to a different class of unusual things than crocodile gods.

In "Prognostiques" (oracles that capitalize on the proliferation of gods) pretended experience and pretended revelation go together. If "superstitious fear of Spirits were taken away, and with it, Prognostiques from Dreams, false Prophecies, and many other things depending thereon, by which, crafty ambitious persons abuse the simple people, men would be much more fitted than they are for civill Obedience" (7-8). In the context of the execution of Charles that is the point. Hobbes has brought philosophy up as a siege engine to batter unorthodoxy, the mad pursuit of which has brought dissidents out of the swamps and woods to depose of a seated king. The individual reception of divine truths vaults commoners above men of inherited rank. Someone like Milton writing his own Christian doctrine and filling in details missing from *Genesis* isn't merely deluded but dangerous. Hobbes' cure for barbarous human nature is sovereign power combining civil and religious authority. That a state of primitive nature and the war of each against every has never existed indicates the extent of his bias. As Elman R. Service (1975) points out, 99% of human history went without formalized government in institutionalized form. Chiefdoms occupied much of that period and at their best functioned judicially in allocating resources and settling disputes. A primitive democracy of councils, excluding women, was an occasional form of government at the small confederacy level. The results were arguably no more contentious than rule by kings and courts, and if one believes Rousseau, Morgan, Engels, Marx, and a few modern anthropologists, somewhat less. The Hobbesian myth is as if no such constraints as schooling, parental care, and cooperative clan behavior existed. Instead, in escaping a barbaric, lawless condition, members of a commonwealth-to-be must surrender their individual prerogatives and accept whatever place the sovereign power assigns them. In actual practice, Hobbes aside, each stage in the consolidation of groups does require the surrender of some prerogatives. But individual, family, clan, and tribal loyalties and residual customs survive mergers that come about by negotiated means. The surrender isn't absolute. Consolidation by conquest is another matter because the conquered have no leverage and cannot demand a Bill of Rights. They are likely to be dispersed as refugees or executed. Deportation, enslavement, and removal were the tactics of emperors on a scale of warfare beyond that of raiders and pirates.

That state-sponsored inhumanity and legal injustice don't appear in Hobbes' commonwealth is hard to explain except as a distortion owed to the fable of uncultured barbarism and the assumption of the fall of man. The political theories of Aristotle and Machiavelli had pointed out weaknesses in unchecked monarchies and oligarchies as well as democracies. Hobbes must have known these, the depredations of tyrants, and in England Sir Thomas More's account of enclosure that turned people off leased property and cast them into destitution. No reader of history or observer of Shakespeare's kings and scheming courts could have been innocent of flawed sovereignty, nor should anyone familiar with the Stuarts have been, or for that matter familiar with the biblical *Kings 1 & 2*. Given the many examples in history, epic, and oracular literature, that is an oversight of magnitude. A strong prince, Machiavelli says, emphasizes not the peace and well being of his subjects but warfare. Ferdinand of Aragon, King of Spain, established himself by attacking Granada right off (*The Prince*, chapter 21). Indeed, a prince has no proper study *other* than war (chapter 14).

Despite Hobbes' distortions and oversights, citations of him in modern anthropology are by no means out of order. His sense of human fallibility and brutality finds support in zoology and anthropology, though under an entirely different natural history. From my standpoint what is missing is attention to illusions at play in collectives however they are assembled and ruled and any significant advance toward a revised cosmology that begins with a heliocentric solar system. He ignores the correction of illusions and of sensory errors through methodical study and testing, which add significantly to what natural philosophy can get reasonably straight. He keeps his distance from two of his most prominent countrymen, fellow monarchists at that, Shakespeare and Bacon, despite endorsing Bacon's empiricism, serving as his secretary, and having him as a patron. The first turns *poesis* to advantage in anatomies of governing orders, particularly kingship. The second uses a combination of common sense and scientific method to improve commonwealths materially. What makes life solitary, poor, nasty, brutish, and short can sometimes be remedied by something other than authoritarian government, such as enough to eat, something less than twelve hours a day in the mines, education, a roof overhead, and trained medical care. That the vast majority had always lived in destitution wasn't the result of anarchy but of economic systems, shortages in resources, and unjust distribution. Unlike his tutor, sponsor, and author of *New Atlantis*, Hobbes isn't interested in the means of material amelioration that political systems and empiricism together might manufacture if they wished, nor is technological and scientific progress a high priority in most kings and their courts. Not much in that line is visible in the 17th century anyway, not until the steam engine and other harnessing of power a century later.

Hobbes' reluctance to take up openly with Bacon could also be due to an instinctive sense that advancements in learning might have a subversive potential of their own. Being freely available, the facts of natural history don't need a privileged hierarchy to be discovered, disseminated, and applied. In that regard access to learning is as dangerous as Puritan saints who read the bible on their own. Both sensory materials and science are independent of government except for sponsored research and freedom from censorship. By establishing priorities and values on the basis of demonstrable merit, they foster a meritocracy of intellect that competes with lineage. Far from illustrating a great chain of being, nature is a leveler in the progression from dust to dust.

John Locke (1632-1704)

Like Hobbes and Hume, Locke bases ideas on sensations, but perhaps because of his reading of Descartes the mind's functions in thinking, doubting, believing, knowing, and willing count for more but not the exclusive demonstration of self existence of *cogito ergo sum*. Mental operations supply understanding with ideas that can't be gotten directly "from things without" (55, 2.1.4). A good many errors can be traced to "Education, Custom, and the constant din of their Party" (252). That infants come into the world with no inborn notions sets aside Socratic preexistence if not all alternate realms. If criteria for truth were inborn, the argument goes, children and idiots would understand philosophic propositions as well as educated adults do. That certain proclivities click in on a biological schedule and that experience sets up structures such as S-V-O relations was too far in the future for anyone at the time to replace pre-existence and a tabula rasa with them. Instead, everything people have in their tool kits they must acquire from either custom or sensations, not from timed biological releases and brain functions or from research monitored by peers. Needless to say, the mechanical level of brain operations was largely guesswork. That it has a quantum mechanics level is as true as it is of everything else, but the advance from that level to classical physics and on to chemistry, molecules, and cells was as far from being known in of Locke's day as it remains in other transitions from quanta to the observable level. Like everyone of his times, Locke was working not just in the shadows but in the complete dark when it came to brain studies. Introspection remained the principle source of information. The biology, chemistry, and equipment needed to go further were missing.

The conflict between convention and what was verifiable in observation crops up in the conflict between revelations and scientific postulates. In Locke revelations remain custom-defined and institutionally administered without apparently the curse of custom. They aren't suddenly inspired moments in which data fall together in a theory that can be tested. Revelation, which Emerson will later define vaguely as "the disclosure of the soul," … "the same fire, vital, consecrating, celestial, which burns until it dissolve all things into the wave and surges of an ocean of light," (245), remains in the keeping of institutions. It is made public through spokesmen in offices gained by taking sacred orders and following specified ceremonies. Unlike precepts in natural history, received enlightenment and doctrine formulated into articles require no verification beyond what they initially received in a burning bush or a stick that turned into a snake. For Locke matter is unknowable at the bottom of the intellectual ladder. Revealed wisdom certain at the top. On that basis he sets out in the fourth book of the *Essay Concerning Human Understanding* to free religion from outbreaks of enthusiasm. Individual contacts with the Holy Spirit are delusory if they are recent.

That orthodox doctrine agrees with reason is Locke's solution. However, since some acceptable beliefs also defy reason, an additional guarantee is needed as to why one suspension of natural law is permissible and another isn't. Unless something extrinsic guarantees it, "every conceit that thoroughly warms our fancies must pass for an inspiration." Hence true faith must be "Assent founded on the highest Reason" (433). That such reason has to be higher than ordinary reopens the prospect of enthusiasm and edges uncomfortably close to the faculty by which prophets claim to make contact with something totally other. Someone as highly placed and respected as Milton claiming that he is writing under the management of a spirit presents a dilemma. Great poems wouldn't be written by madmen, yet this one also wrote prose treatises detailing his own Christian doctrine and defending regicide. The safest alternative is to accept current doctrine, which is essentially what Locke does. He starts there and like the serpent swallowing its tail ends there. Philosophy is again made subservient not to observed data and tested theory but to such myths as a Tower of Babel episode that inflicted different languages on different populations.

Locke's view of rhetoric is equally problematic. Despite his use of it, he rejects it all but categorically. "If we would speak of things as they are, we must allow, that all the art of rhetoric, besides order and clearness, all the artificial and figurative application of words eloquence hath invented, are for nothing else but to insinuate wrong *ideas*... and thereby mislead the judgment, and so indeed are [a] perfect cheat" (202). Several rhetorical devices appear in that well crafted sentence besides those of order and clarity. Since rhetoric, "that powerful instrument of error and deceit, has its established professors, is publicly taught, and has always been had in great reputation" (203), it is only too evident that men "love to deceive, and be deceived." Rhetoric in that sense is mainly the use of adornments and tricks rather than the awarding of names and making S-V-O connections. The divine truth is different in Germany than in England and far different in Saudi Arabia, India, and Japan. By narrowing the scope of rhetoric to its tricks, Locke ignores its most common public effect, the regimentation of populations through early indoctrination. The prevailing judicial, legislative, executive, and religious institutions establish not only laws and ordinances but orthodoxy in other categories and indirectly even in manners and dress. Seemingly unrelated to Locke's distrust of sensations and rhetoric is what looks like a special distrust of magnification (104), which sounds strange coming from a friend of Robert Boyle trained in medicine. In revealing the inner makeup of objects, magnification does suspend normal perception, but it isn't as if normal eyesight became permanently magnified rather than temporarily augmented. In a sense Locke is right, however. The perspective on which conventional beliefs are based is that of the normal human scale and its visible objects. The sun, moon, and stars orbit mankind. Having discovered the microscopically diminutive and the telescopically grand, one can no longer hold human proportions to be a universal standard. What magnification reveals are non human standards, another step beyond Galileo on the way to Einstein's special and general relativity. Increasing the range of vision brings increased alienation from, and wonder at, nature. That the highest magnification possible still doesn't enable us to see photons and neutrinos in action or what dark energy and matter are doing shows its limits without challenging the validity of what it does show.

As to relativity, at the psychological level it is unavoidable. The angle of vision and the accumulated experience of the viewer determine bias. Strictly speaking, objectivity has no point of view. It is all places, times, velocities, temperatures that would be available to omniscience. Looking through the wrong end of binoculars at grazing cattle in Marvell's "Upon Appleton House" shrinks them to fleas as fleas look "In multiplying glasses" (457-462). By Locke's time, telescopes and microscopes had already undermined much of what over the ages had worked its way into credited doctrine and philosophy. The questions they raised were of the kind Milton's Eve asks when she scans the heavens and finds them out of proportion to Eden. She wonders how she and Adam can be what they are led to believe if so much is beyond their range. Curiosity about that is partly what leads her to the fatal Tree of Science, that and the fact that Adam is usually ahead of her in understanding things and the tree offers a chance to catch up. It isn't possible or permitted to transcend where one is placed, but it *is* possible to learn of things divine through archangel teachers. The problem of the number of conflicting claimants to endowed or special revelation Locke solves by accepting the current version of the cosmos, how it was made, and its history. It is given to everyone through the auspices of the reigning institution, which the Restoration had reconnected to the monarchy.

Reliance on revelation leaves Locke little choice but to resort also to biblical taxonomy, beginning with a creation that has established the species once and for all. He is in no position to resolve conflicts between naturalist and other terminologies that leave reason and highest reason strung out between them. In that dilemma he was preceded by Pierre Gassendi (1592-1655) and accompanied not only by Hobbes but by the Cambridge Platonists, latitudinarians, Royal Society members, Hume, Descartes, Leibniz, and Spinoza.

Given nature's extravagance and its abundance of things neither good nor bad, highest reason also has odd things like crocodiles to account for without, in Hobbes' terms, creating special deities for all of them. Neither that nor Satanism is an ideal way to account for reality, but such is the quandary of philosophic discourse that stirs highly unlikely beliefs into empiricism. A good deal rides on keeping the status quo including titles, the ownership of property, manners, and the judicial system. These would lose some of their sanctions if they were cut loose from what the given order was assumed to be and reassigned to social conventions. Those are changeable. Natural laws are not. Given what the natural continuum has generated, no way existed to derive it from an intended design without making the design mischievous or careless. Locke doesn't address that problem directly, but it hovers in the background as it has for nearly everyone who stops to think about it. To account for reality the source must be ingenious, beautiful, whimsical, kindly, monstrous, and cruel plus many other adjectives in thousands of languages. Including *neglectful* means that something else has to account for what resulted from the inattention, or in Leibniz' best of all *possible* worlds *(Essays on the Goodness of God)* find what set the limits.

Natural philosophy has none of these problems, but the 17th and 18th century versions of empiricism were in no position to endorse it while at the same time endorsing providence and sometimes predestination.

Swift's Way with Uplifting Rhetoric

While I'm in the historical vicinity, I want to glance at a quite different combination of realism and reigning misconceptions, this one Jonathan Swift's (1667-1745). Gulliver's travels set before the reader not only a relativity of sizes but radically different cultural practices. The places Gulliver visits are locked into a particular way of seeing things. The dystopian side of *Gulliver's Travels* parodies idealized republics like Plato's, More's, and Bacon's that portray unusual people dedicated to single-minded pursuits. He sets no improved versions of humankind in Gulliver's path other than the rational horses of the fourth book. That simplistic honesty of Gulliver's kind "hath no fence against superior cunning" (2069) reminds us that every illusionist and creator of legends needs a credulous public. Given what Gulliver tells the Brobdingnag king, it is logical to conclude that citizens of other nations, too, not just the English, are members of "the most pernicious race of little odious vermin that nature ever suffered to crawl upon the surface of the earth" (2112).

Among varieties of fraud Swift puts jingoist propaganda near the top and makes the English outstanding in it. The rational horses define a soldier as a Yahoo "hired to kill in cold blood as many of his own species, who have never offended him, as possibly he can" (2151). Gulliver admits to being no stranger to organized mayhem by "cannons, culverins, muskets, carabines, pistols, bullets, powder, swords, bayonets, battles, sieges, retreats, attacks, undermines, countermines, bombardments, sea fights; ships sunk with a thousand men, twenty thousand killed on each side; dying groans, limbs flying in the air, smoke, noise, confusion, trampling to death under horses' feet; flight, pursuit, victory; fields strewed with carcasses left for food to dogs, and wolves, and birds of prey; plundering, stripping, ravishing, burning, and destroying" (2151). He has witnessed the valor of his countrymen blowing up "a hundred enemies at once in a siege, and as many in a ship." With riveting fascination, he has observed "the dead bodies drop down in pieces from the clouds, to the great diversion of all the spectators" (2151).

A rain of mortal pieces isn't the only product of delusions. Governments, science, polite society, and trade keep them company. Science is again impugned broadly, in this case in defense of common sense and the only perspective that comes naturally to people. Mercantile London competes with armies and the mad scientists of Lagado in finding ways to express faddishness, which is part of the wildly imaginative divergence from the norm. In trade the perversions serve "the luxury and intemperance of the males, and the vanity of the females" (2154). Merchants of fashionable goods export necessities in return for "the materials of diseases, folly, and vice." To pay for what they want, buyers "are compelled to seek their livelihood by begging, robbing, stealing, cheating, pimping, forswearing, flattering, suborning, forging, gaming, lying, fawning, hectoring, voting, scribbling, star gazing, poisoning, whoring, canting, libeling, freethinking, and the like" (2154). So much for mercantile capitalism and the fads of consumerism, which are a form of general public delusion. The industrial branch of trade, not yet prominent enough to catch Swift's attention, awaits Dickens and Gaskell. Enthusiasm and curiosity are the current follies. The scientists of Lagado are as much in their grasp as the buying public.

What a journey to strange places and exposure to their cultures accomplishes is a loosening of ties to the particular place and culture of the reader. Relativity in perspective leads to questioning a habitual point of view. The contrast between there and here undermines locked-in ways of customs and habits. The space age was obviously a long way off in Swift's time, but gaining a better perspective on earth and its inhabitants comes of a similar distancing from it. Carl Sagan's video journeys and books (1980, 1994, 1996) and view of the pale blue dot from space (1994) includes everyone of all time, "the aggregate of our joy and suffering, thousands of confident religions, ideologies, and economic doctrines, every hunter and forager, every hero and coward, every creator and destroyer of civilization." All this has come and gone on "a mote of dust suspended in a sunbeam" (8). Extending the range of observations into the macrocosm and microcosm was a major part of modernism's displacement of mankind's traditionally assumed place. That was spatial displacement. In the theory of evolution a comparable temporal displacement was to follow. *Shrinkage* in time and place accompanied *movement* in the natural history to come. Both were behind the bizarre sounds and perspectives of atonal modern music and the art of reorganized body parts in Picasso. Variability in spacetime relativity brings relativity not only in measurements but in concepts.

Thomas Huxley (1825-1895)

The difference Darwin made in extending the planet narrative and bringing contrasting perspectives to time travel was the dominating topic of philosophy after the mid 19th century. It was becoming clear in the debates following *The Origin of the Species* (1859) that because the species were in motion, including mankind, none could be considered privileged, nor could the planet and its inhabitants. The changes couldn't be said to be headed anywhere. Perception of that upended many of the key concepts that remained entrenched in 19th century philosophy. The debates in which Thomas Huxley (1825-1895) participated from 1860 onward indicate what that meant. In three slim treatises entitled "Science and Morals" (1886), "Evolution and Ethics" (1893), and "Evolution and Ethics, Prolegomena" (1894), Huxley put taxonomy and philosophy in the company of evolving kinds. Not much in either ethics or social history could be considered mandated by biology beyond the appetites and the need for sustenance. As world explorers had found, being humanly devised, ethics and social history naturally varied by region if not in quite the extremes of *Gulliver's Travels*.

An early debate mislabeled the Huxley-Wilberforce debate (it included not only Huxley and Bishop Samuel Wilberforce but Benjamin Brodie, Joseph Hooker, and Robert Fitzroy) came too soon after Darwin to explore the ramifications of evolution, but by then it was difficult for a rational person not to reject the tradition reflected in Pope's Pangloss-like phrase "whatever is, is right." Alexander Humboldt in *Kosmos* (1858) had a similar effect, weakened somewhat by his romanticism and tendency to speak of beauty and harmony as if they were universal. Huxley found it inadmissible to classify partial evil as unrecognized good or discord as misunderstood harmony. Nor is chance misunderstood direction. Nature is in perpetual motion and its creatures come in such numbers they can't avoid struggle.

Random luck, too, has something to do with survival, as volcanoes, asteroid strikes, and other deflecting circumstances change living conditions. No one at the time knew how long that had been going on, but geology and the fossil record indicated a good long while. To be adaptable meant only to have survival capacity under specific circumstances. What cultures did for their members, weak or strong ones, was construct shelters and social systems. For efficient operation they included moral codes. In Huxley's words, "Social progress means a checking of the cosmic process at every step and the substitution for it of another, which may be called the ethical process" (81). Ethics and political science take off from natural history but are set against it as much as within it. As distinct from ontology and metaphysics, ethical philosophy sets naturalism aside in building its shelters—artifice meeting nature, roof meeting and repelling sun and rain. Societies establish whatever systems they can within, against, and by means of what the environment provides. That made naturalism the all-encompassing philosophical context and responses to it best directed by pragmatism and realism. The contributing universal patterns of *a priori* reason had their usual functions in managing concepts, but they were no longer a transcendent rationality descending into objects.

One of the oddities about going modern was the way doing so also went ancient, past recorded civilizations into prehistory based on a geography that went back to before the continents had moved into their current locations. The expansion in numbers, times, and distances wasn't bad news or good news. It was merely factual rather than imaginary. Modern historiography tends not to appreciate that fully. It and historical novels and dramas tend to play along with the solemnity and ceremonial dignity. The many dramatizations of Tolstoy's *War and Peace,"* the French Revolution, the American Civil War, the wars of the 20th century, and comparable reenactments capitalize on the drama without providing what Ionesco, Kafka, and

Berthold Breck in the *Threepenny Opera* and *Mother Courage* do by way of a modernist perspective.

Chapter Eight

Empiricism's Razing and Rebuilding

Starting Over

Period names like *Renaissance, Enlightenment,* and *Modernism* are usually reserved for systemic economic and social changes such as feudalism into capitalism. What raises modernism and postmodernism to that level isn't a single factor but a battery of changes that take in not only much of an increasingly visible microcosm and macrocosm but social, technological, and political revolutions. Twentieth century mechanization and eventually electronics and robot labor were the industrial and communications side of that unparalleled overhaul. In philosophy and science, basic questions about existence raised by the Babylonians, Egyptians, and Greeks had to be reviewed more critically under the auspices of empiricism and its vastly improved instruments. As Dominique Lecourt remarks in introducing Evry Schatzman's *Our Expanding Universe* (1992), the universe speeding outward resuscitates "the cosmological concerns of antiquity" (9). By that stage the review of everything from top to bottom was doing as much replacing as repairing and retrofitting.

Empiricism's Razing and Rebuilding

In the early stages of the decision by a few to honor the material evidence wherever it leads, no one foresaw how much would eventually have to be discarded, but where science and philosophy might lead was hinted as early as the fossil studies of Konrad Gessner (1516-1565), not as famously as in the astronomy of Copernicus and Galileo but potentially as unsettling. The fossils contained species no longer around and placed evidence of sea creatures on elevated land. The only technology needed to extract the evidence was a hammer and chisel. Newtonian physics followed Gessner, Copernicus, and Galileo in removing anything not material and visible from natural laws. Up to the 17th century, answering basic questions about natural history was handicapped by the inability to see small and distant. The main advance in technology before electronic devices was that of magnification, but mathematical tools had to be improved also. Tycho Brahe, Johannes Kepler, John Napier, Jost Bürgi, Leibniz, Newton, and Descartes shared in that task. Newton needed mathematical tools capable of defining complex relations such as the gravitational pull of masses on one another at varying distances. In going relatively small more than what the members of the Royal Society had personally seen in not very powerful microscopes came to them from Holland in the letters of Antonie van Leeuwenhoek (1632-1723), the prodigious maker and user of magnifying devices. He found hordes of living small things swarming on his teeth and hands and emerging from black pepper left in water (Dunn, 2009, 52-54). That was startling enough even without anyone as yet able to see molecules.

Like the distrust of the senses and science in Hume, Hobbes, Locke, and Swift, radical skepticism of the Cartesian kind tended to ignore the accuracy of detection enabled by instruments and the reinforcement of one area by another. During those two centuries of growing empiricism selections of evidence allowed even some naturalists to continue assuming that nature was governed by something living, powerful, supremely knowing, and having good intentions. Moral philosophy based on that assumption reached a peak of illogic in T. R. Malthus' *An Essay on the Principle of Population*

(1798), which devoted two chapters (18-19) to repeating that God works betterment of mind through trials of the body, assuming apparently that brains are under the management not of chemistry but a divine being: "The idea that the impressions and excitements of this world are the instruments with which the Supreme Being forms matter into mind, and that the necessity of constant exertion to avoid evil and to pursue good is the principal spring of these impressions and excitements, seems to smooth many of the difficulties that occur in a contemplation of human life, and appears to me to give a satisfactory reason for the existence of natural and moral evil" (123). That most of the animal kingdom suffers with no appreciable mental advancement apparently never occurred to Malthus. Yet it is "to urge man to further the gracious designs of Providence by the full cultivation of the earth" that "it has been ordained that population should increase much faster than food." That produces partial evil but "a great overbalance of good" (114), which is as if to say that famine is justified by more acres of wheat being planted, about equivalent to the lion that eats the lamb inspiring a higher fence. Accepting the evidence but explaining it away requires finding profit in children dying of small pox and Mt. Vesuvius destroying a downslope town. Modernism and postmodernism had their own reservations about the positivist branch of empiricism. More was eventually thrown into doubt than the accumulation of revisions in biology, astronomy, and physics such as black holes in space and quantum uncertainty, and as we've seen discourse in general was included in the movement, especially in the branch that became known as deconstruction. As Marvin Harris in *Theories of Culture in Postmodern Times* (1999) summarizes, by focusing on discourse postmodernism questioned representations everywhere: "Everything wrong with society is no longer to be explained by the mode of production, but rather by the mode of discourse; and the production of knowledge is seen as more important than the production of goods or services" (156). That was not a position taken by everyone, and those who took it might have changed their minds if goods and serves had actually been cut off. As it was, inventories, one of the earliest and most persistent forms of discourse, could in fact list precisely itemized quantities. Their names and measurements were conventions corresponded to real things.

The most crucial of several cosmological revisions came with finding galaxies separating at a velocity indicated by the red shift. Georges Lemaître in 1927 saw the same thing Edwin Hubble did. By then it had become clear what natural history and natural philosophy now needed to do in what amounted to the decisive phase of modernism in its natural history and natural philosophy branches:

- They were to fill in a connected history under constants and invariables from the near the beginning to the present and project the long-range based on the same invariables. A combination of the 'Doppler' effect red shift, the timing of pulsations in cepheids, and the correspondence between the brightness and the distance of stars produced new readings of distance and size, still being refined in calculations of the age of the universe.
- They were to determine the range and force of the cohesive and repelling forces.
- Accounting for nature's heterogeneity needed to start at the bottom with particles where starting there was possible. By working up through composite levels, physics was to find how atoms came together and functioned and, combined with thermodynamics, how nucleosynthesis cooked up heavier atomic weights.

None of these undertakings or the empiricism dating back to Galileo found any function for outside interventions, but to my knowledge the first attempt to construct a comprehensive mechanical model of natural history entirely free of intervening spirits was *Kosmos* by Alexander Humboldt (Friedrich Wilhelm Heinrich Alexander von Humboldt), published in several volumes between 1845 and posthumously in 1862.

Removing spirits, demons, and intent eventually found a parallel in modernist *poesis* attributing supreme fictions to poets rather than to prophets. Few in the earlier 20th century realized what that translation meant more clearly than Wallace Stevens: "The author's work suggests the possibility of a supreme fiction, recognized as a fiction, in which men could propose to themselves a fulfillment. In the creation of any such fiction, poetry would have a vital significance" (*Letters*, 820). That was one phase of modernist poetry. What prophets had been doing all along was exactly that minus the "recognized as a fiction" clause. The sibyls, soothsayers, oracles, and prophets of the ancients were attuned to their times, not to the wavelengths of spirits. When Marduk supplanted Enlil and Tiamat, he put divine insight in the service of different authorities. When in Christianity Yahweh becomes the Christian God the Father, his former chosen people became no longer chosen. What oracles report not only arises from the times but refers back to them as proactive instruments of persuasion. In circuitry that is no different from a T. S. Eliot or Stevens responding to the earlier 20th century and writing for it. The empiricist contribution over the four centuries following Copernicus and Galileo and coming to a comprehensive cosmos in Humboldt is to allow nothing to break in from another circuit as it had been doing in common assumptions for several millennia. Neither science nor natural philosophy was quite ready to stand on its own until it broke free of intrusions, intentional design, and moral philosophy. They needed each other but not the rest.

That kind of intellectual upheaval isn't accomplished overnight. Some of the turmoil of the 20th century and continued religion inspired terrorism of the 21st we can attribute to supreme fictions *not* being recognized as such. That remains a source of science denial and of normal sensory perception, which never makes convincing contact with any of the spirits.

Rereading the Bible

Probably no field of study outside of astronomy, cosmology, and physics has as much need to adjust to modernism as biblical studies did in the earlier 20th century. Being based on belief and having institutions dependent on them they were not well equipped to do so. What instruments of improved vision could potentially do to them had been clear in the displacement of the planet and its people from the center of the universe, a 17th century shock to the contemporary church. If Galileo and his newfangled sky scanning device were right the Bible couldn't be.

Parallel to the retreat in scientific and intellectual circles from moral universe management and magic was a biblical criticism that placed texts in social and historical settings and read them more as literature and rhetoric than as revelation. The texts thereby became more embedded in their times and regions. Once it was clear that the Hebrew anthology contained folklore elements akin to those of Mesopotamian predecessors, attempts got underway to distinguish fictional episodes from chronicle history. Something of Philo returned in the allegorizing of what couldn't be taken literally. Ways to translate mythic figures into natural phenomena became the business of comparative literature. In that context Tiamat translates into the sea as the source of life. When Marduk slices her in two that apparent brutality isn't a grizzly murder but a way to divide land from sea, a division reflected also in *Genesis* where it comes from a voice command rather than a sword.

One result of the Bible/Babel movement at the end of the 19th and beginning of the 20th century was to find polytheistic animations less alien to Judaism than had been assumed. Gunkel's original adversary in the debate, Franz Delitzsch (1850-1922), was an Assyriologist who pointed out Babylonian precedents for Hebrew tales and ritual practices. One reaction to that was to separate spiritual devotion not only from Mesopotamia and Egypt but from some Hebrew chronicle materials themselves. Linking the splendor of

Babylon to the Hebrew religious spirit was repugnant to a way of thinking that had derived sacred writings from divine inspiration. The Yahweh cult based its hostility to Babylon on the complete antipathy between worshipers of Jehovah and Baal. The Holy Spirit had to be detached from narrative details in much the way dualism separates spirit from natural history. The individual parallel is the detaching of a soul from experience and of mind from brain, carried to a neoPlatonist extreme as we've seen in Marvell's "On a Drop of Dew," not an unusual way to avoid the implications of bodily existence.

Concerning biblical chronicle materials and incidents, despite what the texts actually said and whatever their sources might have been, Gunkel found a religious spirit arising from the text. In *The Influence of Babylon on the Religion of Israel* (1902-1903) his defense of that is adamant. Delitzsch's lectures featured reproductions of Babylonian pomp, processions of women slaves, idols, Sargon-empire architecture, lion hunting expeditions, displays of costume and weaponry, and warlord postures. To Gunkel these were popularizing stunts ignorant of "the *science* of religion," in short "*dilettanteism*" in full bloom. To literary criticism, similar stories appearing in more than one culture suggests that they were conventions, like the story of the infant Moses floating down the Nile in a basket (Exodus 2:1-10). To those who take up the claims of the text itself, the difference between Babylon and Judea remains much as the Hebrew scribes always said and as the tradition has maintained, also as movies, novels, and operas still present it uncritically.

What Gunkel objected to wasn't the brutality of the Israelite conquests but the compromising of the religious spirit. Scholars who wouldn't ordinarily have chosen an exclusively inner contact with it and a separation of natural and spiritual realms felt they had to do so given the chronicle details and foreign elements. Even that, however, wasn't entirely new, nor did it cease with the new criticism. Pauline exegesis, too, was a cleansing operation with a Platonist element that preferred ethereal to material things. Topographic irregularities become irrelevant, too, along with most of what is acquired from experience. Rather than being members of a nation speaking to its people, the prophets in Gunkel were tongues of flame speaking to posterity. That was in spite of the fact that in the text they applied Yahweh and the armies of Israel to real nations. To readings based in social realism, Yahweh followers lived in a historical world with rival religions, apostasy, heresy, and hostile military powers. If the prophets and their texts had anything ethereal about them it doesn't show in their exhortations, especially when in *Deuteronomy* style they detail the discomfiture of Israel's enemies. Gunkel's portrait of the "indignant Amos, the majestic Isaiah, the deep and tender Jeremiah" edit out that nationalist element. The difficulty of doing that logically shows to this day in commentary on the contrast between new and old testament deities, where an international evangelical movement must avoid nationalist militancy. Punishment of those who remain outside the chosen circle is placed out of sight in a realm of its own in every way the opposite of the spirit realm. For them the former Judaic war god becomes an eternal torturer. What was war in the texts becomes quite literally hell.

A further drawback to an internalized spirit was the dubious authenticity of any given prophet. In the wake of the Puritan rebellion as we've seen, Hobbes and Locke found civil turmoil being generated by what they considered incredible claims to divine insight. Locke was especially anxious to separate authentic revelation from delusion. The turmoil of the Reformation wars and the English civil war made that all the more urgent, but it had often been an issue in both testaments and in Europe's Protestant rebellions. The Reformation wars were essentially over which institution was genuine and which not. Insofar as they, too, were nationalist oriented, the Hebrew example remained at work. Extraordinary revelations to one party often bring attacks on rivals. That has less to do with comparative credibility than with ethnicity. To gain followers, the leaders of a new sect must discredit their predecessors. In the case of Paul and Peter, gentiles and Jews called for different evangelical campaigns. Persecutions continued against the early Christians, in Catholics against Protestants and Protestants against Catholics, and against and by the Mormons. Additional prophets continue to come forth. Among sects that have become momentarily notorious in modern times were the Rajneeshees of California and Oregon who in 1984 poisoned voters of Wasco County in an attempt to control an election. That was an unusual but not a unique sect-generated militancy, accompanied in that case by a fleet of luxury cars and semi-automatic weapons if not the Ar-15s popular among terrorists. We have sampled the rant of Abdullah Yusuf Azzam, co-founder of the Al Qaeda movement with Osama bin Laden, in that case not a new movement.

Similar professions of revelation turn up in less well-known documents. The Mesha inscription, called the Mesha Stele or Moabite Stone (9th century BCE), erected to boast of conquests, tells of a campaign directed against the Hebrews, as Christian and Muslim campaigns will be as well. Translations vary but the main events are clear enough: "I fought against the city and took it," the inscription claims. "And I slew all the people and the city became the property of Kemosh and Moab. And I carried from there the altar and Kemosh said to me, 'Go! Seize Nebo against Israel.' So I proceeded by night and fought with it from the crack of dawn to midday, and I took it and I slew all of them, seven thousand men and boys, and women and maidens [spared in one translation] because I had dedicated it [them] to Ashtar Kemosh. I took the vessels of Yahweh, and I dragged them before Kemosh." In a losing battle Mesha makes a display of making his eldest son a burnt offering, at which point the horrified allied forces are said to have forsaken Israel's attack and returned home.

Derision continues to be part of the rhetoric on both sides, as it is in the Mesha inscription and in Hebrew writings all the way forward to the Dead Sea scrolls. Belief in a righteous cause and disdain of outsiders reinforce one another. In the epics and myths it is voiced by both gods and human combatants. The Assyrian Sargon and Rimush stelae, the Shalmaneser obelisk, the Iran stele concerning the despot Tiglath-pileser, and the chronicles of the 8th century Sargon II tell of campaigns similar to Mesha's. Epic taunts crop up in accounts of Persian, Greek, Carthaginian, and Roman campaigns and on into the Crusades and Reformation wars. The gods range from Enlil, Marduk, Shamash, Ashur, and Kemosh to the Greek and Roman pantheons. The literature takes in epic, tragedy, lyric, mythological forms, and what pass as chronicles. Delusions of grandeur underlie the taunts. Since someone must lose for someone else to gain, the presiding overseeing power necessarily abandons one side or the other. Ovid's *Metamorphoses* (1993) is characteristic of worship gone sour, in that case not for any particular reason:

> Wherever I would turn my gaze, men lay
> along the ground, as acorns when one shakes
> the oak or rotten apples when boughs sway.
> You see the temple with its long stairway;
> that shrine is Jupiter's. Who did not
> bring unto those altars useless offerings?
> How often did a husband, even as
> he prayed for his dear wife, a father as
> he pleaded on his son's behalf, meet death
> before those cruel altars—with a bit
> of unused incense still within his hand! (Mandelbaum, 233; 7.583 ff.)

Thus the common pattern: exhortation fueled by derision in one phase, lamentation following.

John Van Seters' *In Search of History* (1983) underscores the writing of enhanced annals among the Hittites, Egyptians, and Israelites, but again the Hebrew anthology has the best known and most numerous examples. Fortunately for textual interpretations of Julius Wellhausen's comparison-and-contrast kind, *Kings* and *Chronicles* offer duplicate versions of several episodes conceived differently according to the rhetorical circumstances. From the differences between the *Kings* and *Chronicles* we can see adjustments to the times if not the social and political situations in any detail. Wellhausen in 1878, the first and one of the most thorough to put the two texts together, came to the conclusion that "the old picture is retouched in such a wise that all dark and repulsive features are removed, and their place taken by new and brilliant bits of colour not in the style of the original but in the taste of the author's period– priests and Levites and fire from heaven, and the fulfilment of all righteousness of the law… the material of tradition seems broken up in an extraneous medium, the spirit of post- exilian Judaism" (187). Post Babylonian Judaism is quite different from pre exile Judaism.

For earlier narratives in the Mesopotamian background we no longer have the names of the poets and chroniclers. The sources and later redacting authors of biblical texts remain unknown as well, but internal evidence contains clues that scholars have had some success in deciphering. Placing the texts in their probable times and circumstances suggests again that the intent was to rally national support. The recruitment was thus simultaneously into the cult and into the military. That warlord conflicts among early city and territorial states presumed supernatural participation as well suggests that what the inscriptions and texts reveal was a general rhetorical practice that William J. Hamblin (2006) finds typical of the Near East (24-34) in general, including the gods of Akkadia (2334-2190 BCE), Anatolia (11,000-5500), Syria (10,000-4000), Elam and Iran (7000-3400), Egypt (several dynasties back to 3500), and Canaan. Over a span of three millennia that area and westward in the Mediterranean saw the rise and fall of gods along with monarchies among the Sumerians, Akkadian, Chaldeans, Assyrians, Babylonians, Egyptians (several times), the Israelites (several times), the Medians, Persians, Hittites (in Anatolia), Phoenicians and their colonists in Carthage, Mycenaeans, Kessites, Amorites, Grecians, Romans, and several lesser city states and island polities like the Melians and Corcyreans.

I cited Mesopotamian parallels earlier in Sargon's (2334-2279) claiming to have made his conquests with the help of Anun, Ishtar, and the mace god Ilaba. His son Rimush (2278-2270) rid himself of unwanted populations, replacing them with rewarded Akkadians.

Naram-Sin (2255-2215) declared himself to be god of Akkad and tells of putting down rebellions and displacing populations in that role. Archaeology has turned up less legend-checking detail about civilizations in Sub-Sahara Africa and areas of Asia and Europe, but historians tell us that the Balkans, Russians, Norse, Franks, Aztecs, Koreans, Mongols, Huns, Japanese, east Indians, and Chinese likewise mounted invasions behind mysterious help speaking only to selected visionaries. It wasn't any particular beliefs that secular modernism replaced but all of them as a recurrent psychological pattern. The evidence came from several directions and mutually supportive disciplines. Typical of fundamentalist returns to any one of them is either a general science denial, ignoring the evidence, or concentration on one or another discipline without attention to the cross referencing of several as geology and evolutionary biology collaborate in the theory of evolution.

Another oddity of animations varying by time and place is that they are said to demand devotion from entire nations and yet have chosen to achieve it only through a prophet or two rather than making a general public appearance. The only testimony available to the public comes from the oracles themselves, and those of one region differ from those of another. Elman R. Service (1975) pushes the connection between centralized rule and ritual a step further back than state formation, finding a "chiefly line" likely "to become a priestly line . . . interceding with its ancestral gods in favor of the society" (78). Not every primitive group would have been organized in the same way, but at the tribal level chiefdoms "ethnologically seem to be typically, perhaps universally, theocracies." In Service's concept of typical tribal chieftains, spiritual matters reinforced their authority as they later did that of warrior kings in city and territorial states. It is conceivable that they played a role also in the qualified democracy of council-governed clans and hamlets. That they still do in congregations and democratic states with a separated church and state shows in national anthems, pledges of allegiance, and coin of the realm inscribed with token piety.

Archaeology treats social phenomena as phases in human development best explained in collective psychological terms. What Hamblin finds true of Egypt goes for other nations and empires as well: warfare "was not perceived as merely the struggle between mortals; it was theomachy–the war of the gods" (353, 412, 415). Personal ambition, desire for land and plunder, and other common motives did their customary work behind lofty fronts, but as Keeley points out (1996), "the question of what motivates an individual or group to engage in warfare is a vexing one" (113). The causes are variable because minds are complex and many of them in concert even more so. Subjugation, a branch of hegemony, is itself a mixed bag, "a rubric that subsumes disparate goals of defense, revenge, economic, and territorial gain" (116). No doubt some of the war god beliefs were sincere, but objectively speaking it remains improbable in the extreme that supernatural animations actually take a personal interest in battlefield outcomes despite the popularity of the 'god's on our side' syndrome.

Finding claims for sanctity to be psychological puts the cleansing operation of imperialist campaigns in the bitter tone of Mark Twain's *Letters from Earth*: "Oh Lord, our Father, our young patriots, idols of our hearts, go forth to battle. Be Thou near them! With them–in spirit–we also go forth from the sweet peace of our beloved firesides to smite the foe. O Lord, our God, help us to tear their soldiers to bloody shreds with our shells. Help us to cover their smiling fields with the pale forms of their patriot dead. Help us to drown the thunder of the guns with the shrieks of their wounded, writhing in pain." And so on through much more that is remindful of Gulliver's lists in answer to what is missing from podium rhetoric. Glorification begs for satirical response as the pompous figure in the Commedia dell'arte generated the mimicry of the one who shadowed him. A trace of that tone turns up unexpectedly in the mockery comedy of Milton's Satan launching missiles into the angelic host in imitation of declarations from the Father and Son. The highest of dialogues becomes the lowest form of sarcastic mockery.

It seems that few if any urban civilizations have been without lofty myths put to nationalist or personal use. By about 10,000 years ago the Neolithic may have already acquired oral myths that later surfaced in written forms in Uruk, Nineveh, Akkadia, Assyria, and in Kassite and Babylonian kingdoms in the south. Separate realms for gods and spirits were assumed in Egypt, Anatolia, and a few centuries later in Canaan. Service surveys tribes in Africa (Zulu, Uganda, and the Congo), among the Cherokee nations, in Polynesia, Mesoamerica, Peru, Mesopotamia, Egypt, Indus River Valley, and China, and glances at others in passing and finds none of them lacking in ceremonies dedicated to imaginary powers. Those powers weren't entirely aloof but were sporadically active in nature and human affairs. It is unlikely that an incident-free, non-material, faith-only alternative like Gunkel's existed, but if it did it wouldn't have etched icons in durable materials. Written Mesopotamian myths put kings and priests in contact with gods through rituals, sign reading, and sometimes the sacred violence that Paul Kahn chronicles (2008) and Mesha's Moab Stone and other stelae illustrate. The multiple examples of war-minded figures before and after the Hebrews one argues for the Delitzsch reading of Assyrian influence on Israel and against Gunkel's etherealizing of the text.

Even so we can't assume uniformity of beliefs and devotional practice throughout a national population. Once civilizations have reached an advanced occupational specialty stage and acquired cooks, ranks of warriors, clothing makers, weapons makers, farmers, courtiers, diplomats, priests, and shoemakers, and adversarial groups, they would likely have generated some who were skeptical of national myths and rituals. Hallpike concludes in that regard that the development of writing, calendrical science, and mathematics existed alongside, and was compatible with, a mythical mentality, which wasn't "significantly different from that of tribal society" (235). Assuming that what the case is now has precedents, not only people of common sense but the wise as well probably kept some distance from overly zealous believers. That art and texts to that effect are lacking isn't conclusive proof that realism itself was missing. What does exist are diatribes against non believers and those who chose something to believe contrary to what the authors did. To the latter enough outsiders existed to pose a threat. Where patronage was required, support would have gone to a given orthodoxy rather than to dissidents. Also, art is more devoted to the extraordinary and spectacular than to the humdrum. No one writes epics about happy households or the barley harvest. Lyrics, too, tend to be pitched higher than that. It remains *poesis* that calls on muses, gods, and goddesses and groupthink that spreads and perpetuates belief in them.

As the Moab stele illustrates, what did get noticed were only a few highlights until parchments and scrolls allowed more elaboration. None of the written records can be expected to treat dissent with sympathy any more than moderation writes epics. Whether the account is put as a chronicle or as an epic poem, the literature favors the invaders where they have succeeded. Had there actually been a conquest of Canaan we wouldn't expect to have an account of it from the Canaanites, most of whom the texts say were killed. Alexander gets good reports from Greek and Roman historians who offer mainly highlights. Carthage gets bad reports from Roman historians. Napoleon gets mixed reports, Hitler obviously quite different ones than party propaganda generated at the time. For imperialistic nations over the next many centuries, the exodus legend became a prototype of conquest as divinely ordained national fulfillment. The point of view of the victims is usually missing until the invention of the printing press. The psychology of invaders is much the same whatever their banners say.

Whether predominantly uniform or not behind sanctified movements, populations do have to be committed to a tribe or nation in sufficient numbers to make it work even for peaceful, non-expansive purposes. The parties whom Nehemiah lists by genealogy and by tribe collaborated in the rebuilding of Jerusalem even without civil authority to gather a labor force. What they had was the preaching of Ezra and the industry of Nehemiah. That, to repeat, is again evidence of lofty myths raising as well as razing civilizations. It is one of the recurrent energizing social forces.

Texts and Excavations

Gauging discrepancies between fact and fiction and their proportions when they are mixed ranges from difficult to impossible in anything very ancient, but artifacts that have not been tampered with for millennia are more reliable than texts that have been. That use of durable materials is part of the record that can last millions of years and potentially even longer in hard scrabble in space. Even when carved in stone, texts are another matter. Did the Assyrian Assurnasirpal II (883-859) actually flay and skin people, impale them on pillars, cut off their limbs, noses, and ears, put out their eyes, and decapitate them in the name of Ashur (*Assurnasirpal* means *Ashur is guardian of the heir*)? Perhaps, but in the context of bronze and iron age propaganda, we would be naive to think so without supporting durable evidence. How much of the accounts of conquests by Shalmaneser III (858-824), Tiglath–pileser III (745-727), Sennacharib, Nebuchadnezzar, and the rest of the Babylonian, Assyrian, Persian, and Greek legends can we credit literally? Again, clearly not everything and perhaps not much. No reliable historians or reporters were around to chronicle the events. We do, however, have the history of wars between 'take no prisoners' adversaries and know that torture and mutilation accompanied both tribal and national illusions and delusions. The fact checking falls to area specialists, who for over a century have been unearthing and learning to read what evidence there is.

Although inscriptions and texts are more likely to be misleading, unearthed materials, too, have to be interpreted. Burial sites and depictions of battles such as those of Delitzsch's Assyrian figures depict weapons designed exclusively for war. In surveying the span from Australopithecus to Neanderthalensis, Arther Ferrill (1985) finds similar evidence that spears were used in battle as early as 70,000 years ago. Pre Neolithic bows, slings, daggers, and maces add to that: "Neolithic cave paintings clearly reveal their use against men as well as animals" (19). That extends the range of warfare to the era of kinship bands thousands of years before Ashur encouraged Assurnasirpal to torture his victims. Steven A. LeBlanc in *Constant Battles* (2003) agrees.

It isn't far from the Anu, Ishtar, Marduk, and Ashur to their equivalent in monotheism. Monsters such as Zu and Namtar collected into Lucifer and into the gods themselves, in the petulance of Zeus, for instance. Despite their diversity, the former populations of contemporary Balkans, Macedonia, Montenegro, Romania, Serbia, Bulgaria–the battlegrounds of Byzantine, Bulgarian, Serbian, and Ottoman empires–continued to apply similar binding powers to their national agendas. The motives for drawing on myths have to be speculative, but behind some claims for supernatural input would have been a desire to be more assured of victory. Before an arm hurls a javelin or throws a grenade, the mind must be prepared to do so, and for an organized army to go forth many minds must be in synch. Marching formations and hymns of war are two of the recurrent coordinating devices. A contrived high cause is another. A chorus of Judean slaves in Verdi's *Nabbuco*, based on *Jeremiah* and *Daniel*, is an operatic version of the high-spirited unanimity. Removing the war god from it and laying out the articles of a new Italian republic, the contemporary parallel, would require writing a constitution rather than singing a hymn, though even constitutions tend to have high myth preambles cited many times thereafter in variants of "we hold these truths to be self evident."

The numbers put into the field needn't always be large. Hallpike (2008) contends that "Comparatively small professional armies could conquer many times their number of tribal opponents because of their superior discipline and organization…. The development of large organized battles… made it possible for single, decisive battles to change the course of history" (166). As Keeley points out, however, that common assumption isn't universally true and contrary instances aren't rare. Diseases that get through immune systems are more devastating than weapons. In any case where ancient conquests have succeeded that is best proved not by texts but by changes of artifacts in the layers of excavations. Where these verify what texts say archaeologists can be sure a change of regime and culture took place. Where weapons embedded in bones are found in the ruins they can be fairly sure it came about by conquest. Where a text says such a conquest took place and the artifacts show nothing of it, specialists in the area can be fairly sure that what looks like history in the text is mostly not. The probably non existent Canaan campaign was incorporated into the Hebrew anthology centuries later as legendary material. Reasons for the reign of Josiah to have done that remain tentative, but in some interpretations they have Josiah's reform movement and the desire to reunite the North/South divided kingdom behind them.

Evidence in Rock

These have been only samples of the myths that survive in the age of empiricism. What they set aside is generally not only more substantial but many times more formidable. The record in rock, one of the nearer repositories of evidence, assists not only the writing of amended history but comprehensive prehistory. As a reminder, here is reasonably well established natural history information from mutually reinforcing sources, one of them from a rock planet, to set across from stories that had been superimposed on the planet history. The Milky Way and the sun have been evolving 750,000 times longer than what typical myths of the universe formation propose. Earth history has tolled off over 234 billion of the seven day cycles of *Genesis*, all but the last few passing before anyone thought to call Son-dayes the Out Courts of glory. Rather than the harmony of the spheres, the galaxies and galaxy clusters contain mixtures of geometric spheres, including satellites in slightly wobbly elliptical orbits, and randomness in materials that gravity never managed to pull into spherical heated masses. The material a supernova sheds goes into new stars and satellites and leaves dispersed fields of debris. One recently discovered addition to the chaos from studies made in China and NASA is the way large black holes swallow stars. Stellar tidal disruption is the product of variants in a black hole's gravitational pull that 'spaghetti' a star caught in the field of influence. That can be inferred from flares of energy and light. Like other celestial phenomena, the awesome beauty of such a thing depends on the observer being a safe distance away and having the right equipment to see it. In magnitude, pillars of discarded material are many billions of times more extensive than the pillars of broken rock thrust up from earth's surface.

In the company of data of that kind, which astronomers and geologists expand into much more, the inflation of high myth rhetoric shows clearly. Humans are an ego centered race whose inventive brain doesn't always distinguish between the subjective and the objective. The appeal of the myths is partly the downsizing of natural history to put mankind in a better proportioned setting. Since what falls within the sensory range comes from what is outside it, we too are made of particles. What the senses take in has as long a mechanical history as the stars do. What landscapes and topography reveal to naturalists is too large and sometimes too confused to focus too exclusively on a nation or even on mankind as a whole.

Back down to earth set in this modernist context we find a naturalist prose as impressive in its own right as the literature of epics, myths, and oracular sermons. Consider Peattie (1938) writing about a grove on the great plains. A bounded grove is parallel to the essay itself in isolating a defined topic amidst an indefinite extent. The grove and the essay together establish points of reference as if to say, 'give me all nature for the far range and a single locality as a focal point. Then I'm at home':

> If you cross the prairies by the train, it seems too far from here to there; that is because in the flying coach you are never anywhere. But that woman who waved at you from the porch of the white frame house under the silver poplars is in the geometric center of a circle as wide as the visible ends of the earth. She has had time, living there, to learn how mighty is distance seen horizontally. Some people only love a hill; they like their views prettily framed for them. For such, mountains are excessive, and plains give them agoraphobia. But if I cannot have mountains, give me a plain where there are a hundred and eighty degree of sky arc. For my peace, my habitation, and my heritage, give me an island grove upon that plain (5-6).

In standing apart, the island grove provides a frame of reference like the margins and first and last lines of a sonnet or the wooden frame of a picture. For an earthling an unanchored, free floating self in such an immensity is uncomfortable. Nature provides the stage, language the warehouse of materials out of which to build a topic. Nature's more usual boundaries are rough, like jagged shorelines, and its terrestrial features vary from nearly flat to jagged. Nature's buildup comes in variable sizes and can be as complex and beautiful as a coral reef. Roughness and vast extent aren't objectionable in themselves, only obtrusive in earth histories and cosmologies that deny them or try to moralize them.

For Aldo Leopold and John Burroughs, too, meaning arises from an essay's concept of a focal area or time period. Leopold (1949) frames a morning scene:

> To arrive too early in the marsh is an adventure in pure listening; the ear roams at will among the noises of the night, without let or hindrance from hand or eye when a flock of bluebills, pitching pondward, tears the dark silk of heaven in one long rending nose-dive, you catch your

breath at the sound, but there is nothing to see except stars It would seem as if the sun were responsible for the daily retreat of reticence from the world. At any rate, by the time the mists are white over the lowlands, every rooster is bragging *ad lib*, and every corn shock is pretending to be twice as tall as any corn that ever grew. By sun-up every squirrel is exaggerating some fancied indignity to his person, and every jaw proclaiming with false emotion about suppositious dangers to society, at this very moment discovered by him (61).

The pleasure of half light is that of surmise growing into certainty. The evidence in this case is ephemeral and uncertain, but dawning intelligence answers to dawning vision. As Peattie (1938) remarks concerning a similar in-between state, "Curiosity is not an instinct; it is a growing tip of the intelligence, it is a divinity upon the darkling animal mind, like the upward yearning of a vine toward sunlight" (86). The rooster and squirrel, guilty of self aggrandizement, are all too human, they too being self oriented first and objective second. Suppositions have their appeal, as in the unseen bluegills tearing "the dark silk of heaven in one long rending nose-dive." Inference fills in where daylight isn't yet full. In effect by singling out an object, perception isolates it and finds a momentary focus.

Such location is both satisfying and temporary. We can't hold onto it, merely come upon comparable realizations from time to time. Once a history is sketched in, more of the greater context comes into view. In Burroughs (2001) a given location manifests a comprehensive history: "Your lawn and your meadow are built up of the ruins of the foreworld. The leanness of granite and gneiss has become the fat of the land. What transformation and promotion!–the decrepitude of the hills becoming the strength of the plains, the decay of the heights resulting in the renewal of the valleys" (137). Resulting, too, in the soil of the valley becoming the silt of the river and the sand of the shore. These in turn become something else as the story goes on.

Our particular place and moment can be singled out and assigned a boundary, but it can't be understood without glimpsing considerably more. When we pass from the evidence of archaeology to that of paleontology and geology the framework takes precedence more consciously and quite specifically in how the succession plays out. It is that framework that determines how much of a given comprehensive belief is valid, and it is ruthless in its judgments. The cycles of buildup and breakdown extend out of sight in the past and beyond calculation in the future. Meanwhile unlucky species give way to something else, perhaps better, as we think ourselves superior to reptiles and primates. The particular dinosaur never returns, only by chance its bones reassembled in a museum. Needless to say, the reassembly depends on the function of the parts, not on a concept of what an ideal dinosaur should be. So with the natural continuum generally. It grandfathers nothing in for continuity. The burnout of stars in the Milky Way and its Laniakea Supercluster lies about a trillion years in the future, so David J. Eicher estimates us in *The New Cosmos* (2015). The evaporation of black holes and other finishing phases will depend on whether the visible universe is singular or one of many. The rock evidence at hand takes us only a third of the way back and light evidence close to a glimpse of the beginning.

Myths of the Commonwealth

Chapter Nine
Militant Rhetoric and War Gods

National Identity

Let us keep assuming that none of the sky powers ever did what ancient battle chronicles supposed and some radical movements still claim. The question then becomes what the rhetorical purpose of the claims might have been. One incentive to make them is to get a nation aroused to a pitch to defend itself or to go on the attack. Patriotism is often just such an amplifier, and the gods are notoriously nationalistic, especially the war gods. Along with them and the rewards of public service, stereotypes and demonizing reinforce national level undertakings. At the individual and household level, arousing self defense doesn't require that kind of urging.

But back up a moment and consider a broader spectrum of enabling forces for clans, tribes, city states, and nations. When tool invention and technology are included in a survey, a plausible case can be made and usually is for the mode of production as the tangible factor. The imaginary powers that I've cited in myth, epic, and oracular sermon work only indirectly in the coordinating and motivating of large populations. By indirectly I mean psychologically and through groupthink illusions. The rise of civilizations depended on agriculture and industry and their implements and tools, made of copper and its alloy bronze following the stone age. That was what enabled the gathering of urban populations and organized the workforce in tangible ways. It didn't necessarily require a dynastic ruling elite, a large military, or imperialistic campaigns. The economy enabled these but didn't coordinate and direct them. That fell to dynastic leadership, which is where the added authority of divine sanctions often entered. In brief, the inventors of the metals and tools were the chief enablers from the point of view of production, not those who in the tall stories fought the devils and monsters. What built the early urban areas and their wealth were actual labor, tools, and expertise. What supported the dynasties for millennia were accumulated wealth and illusions. The ancestral heritage based on territory and legends played a motivational role. In Israel the legends stationed demonstrations of divine intrusions in the distant past when seas parted and the sun stood still while a battle continued. Since the wealth gained from imperialism built monumental architecture, war has to be included in the mode of production, but it only transferred wealth. It didn't create it.

The change from hunting and gathering to agriculture and industry, paleontologists and archaeologists concur, was global except for isolated places. It took over because it supported greater populations and generated the economic surplus used to build cities and support armies. While food remained wild, scattered, and seasonal, the bands that pursued it were limited in size. Warfare at the edges of the two systems forced changes in nomadic tribes. When one city was joined to others by conquest or negotiation, territorial and city states became nations. The most familiar example is the nearest one historically, the introduction of farming and manufacture into North American areas of stone age people from the arctic circle south to the US four corners region. The motives in operation were economic where they were tangible, but motivational racial, nationalistic, and mythic ones were often added to them.

The more developed and elaborate the civilization the more detailed and complex its enhanced history and the more systematic its beliefs. These are likely to become ritualistic and elaborate enough to be turned over to professionals. Something like an ecclesiastical hierarchy often accompanies the ruling order or is incorporated into it. The Christian ecclesiastical hierarchy from the 4th century onward mimicked the civil one in multiplying graded ranks in a chain of command. A king in heaven surrounded by ranked angels argued for a monarchy and court. Chronicles, historians, and dramatizations all agree in portraying a social elite concerned foremost with rank. In lineage rivalries over hegemony and succession, both sides claimed sanctions in a higher order, stirring natural and moral philosophy together to the confusion of both. The elite sometimes projected an equivalent hierarchal structure in the animal kingdom and in ranks of angels, thus making the chain of command also a universal chain of being. The question for its human branch was frequently which of several rival claimants was sanctioned, the civil realm equivalent to determining which of several prophets was genuine.

A vertical universe became harder to maintain when the middleclass benefitted from the same spiritual edification as the elite, which in England meant Puritans contesting church, king, and court. Unlike the spurned lower orders of lords-and-serfs monarchies, middle class males were educated and mostly non agricultural. The leisure class found itself on the defensive, not only the cavaliers in England but in most of Europe. The later movements in the French and Bolshevik revolutions assumed national identities free of the paraphernalia of monarchs and tzars while preempting the absolutist language of theology.

As populations grew so did the reach of communication and of travel, and as these expanded so did the range of weapons. The larger the battlefield the more the launching force and payoff at the tip needed to be, and the more extended the reach of the indoctrination that enlisted the combatants. On the whole the terrors of losing are enough in themselves to need no other motives for self defense.

With one exception I agree with Susan Niditch (1993) concerning Israel as an early case in point: "All the attitudes of war we have uncovered in the Hebrew Scriptures are as old as human culture itself and as complex as human thought, linking our earliest ancestors with ourselves and our neighbors' cultures with our own" (155). The exception is the age of those attitudes. When and where calling divine beings in against demonized enemies took root is only as clear as murky historical conditions allow, but it looks as if paranoia, hatred, dedication to inflated causes, greed, and bloodlust came bundled and spread early on. That so many tribes and nations have been peaceful is a triumph of discipline and reason over those easily triggered instincts that reigned for millennia in Mesopotamia, Egypt, and on into the European empires, including the Holy Roman Empire.

The highest of the high ground falls to sacred causes. The earliest imperialism on record made prominent use of them. To Sargon as later to Tiglath-pileser the defeat of enemies was a measure of the love the gods bore them: "Like a storm demon I piled up the corpses of their warriors on the battlefield *and* made their blood flow into the hollows and plains of the mountains. I cut off their heads *and* stacked them like grain piles around their cities. I brought out their booty, property, *and* possessions without number" (Arnold and Beyer, 128). As Keeley points out in *War Before Civilization: the Myth of the Peaceful Savage* (1996), "there is simply no proof that warfare in small-scale societies was a rarer or less serious undertaking than among civilized societies. In general, warfare in prestate societies was both frequent and important. If anything, peace was a scarcer commodity for members of bands, tribes, and chiefdoms than for the average citizen of a civilized state" (39). Group psychology probably wouldn't have changed greatly, a conclusion that finds some support in Azar Gat's compendious *War in Human Civilization* (2006) and from archaeological discoveries of weapons. A top-down chain of command takes charge in the field and often in the society as well.

Once immortals become a common heritage either keeping them around or removing them raises problems. They are as tenacious in holding on to their powers as the hierarchies that administer and use them, as Sargon, Tiglath-pileser, and their heirs did. In Book 10 of *The Aeneid*, Jupiter taunts Juno facetiously concerning who actually determines the outcome of conflict, men or gods: "It must be Venus who sustains the Trojans,/ Not their good right arms in war, their keen/ Combativeness and fortitude" (10.853-55). That Virgil's Juno continues to raise trouble for six books before she capitulates tells us that she is a narrative device as well as a figment of the imagination. She must provoke twelve books worth of Roman ancestral history and look forward to further conquests. Such an overpowering force could end the conflict at any time if the conventional form of the genre and the on-going story of the empire didn't forbid it. In Milton, too, actual history continuing on makes it necessary to withhold any definitive use of supreme power. The final solution recedes as natural and human history advance. Nature can't be replaced by a paradise until the conditions are right, and only God knows the appointed hour.

Founding Stories

Although militant imperialism hasn't always coincided with the ascendance of male figures, Trigger (2003) observes that "wherever the status of women can be traced over time, it appears to have declined rather than improved" (194). That seems to be true of goddesses as well as real populations. It may be a cyclical and cultural matter not a global trend, but it has had prolonged episodes that can be traced to some extent in mythology. Not only in Mesopotamia, Egypt, Persia, and Israel but in Greece, goddesses receded as imperialism and state sponsored piracy took root (and loot). The rise of Marduk coincided roughly with Mesopotamian empires. Something similar seems to have happened in the Persian Achaemenid empire in the mid first millennium, in Minoan civilization under Agamemnon, and in the Athenian displacement of the Mycenaeans. Had Israel been larger and richer in resources, its tales of conquest in the Mosaic past might have inspired imperialism there, too. As it was the destruction of Egypt, Assyria, and Babylon were imaginary visions by those who expected Yahweh to supply the means. In the later prophets and the Dead Sea scrolls, he was reported to have done so in the past and could be counted on to do so again.

Even where shrines to goddesses lasted and scholars find evidence of female consorts, warrior kings, and despots favored militant male divinities. In the text version of the Canaanite campaign, Yahweh ruled supreme but Asherah and Astarte hung around in some areas. Their more prominent presence would have divided a population that the exhortations were designed to consolidate. The mixed state can be seen not only by the consort sometimes attached to Yahweh but by the actual status of women in Israelite social and household history (Carol Meyers, 1988). At a minimum we can say that households away from Jerusalem entertained rituals not as devoted to Yahweh as the central priesthood wanted. The reform movements of the central temple attempted without complete success to root them out. Intermarriages, population movement, and cultural exchanges worked against uniformity as did distance from the center and age-old tribal differences, which became less after the Assyrian conquest and disappearance of the 10 lost tribes. That raises a problem only where the nation is closely identified with a given set of rituals and an enforced orthodoxy. No distinct pattern applicable to every nation emerges from that, but current rulers in league with clerics and scribes typically connected rituals and beliefs to the earlier founding of the nation and called for national unity to prevail over populations at a distance from the urban center.

In Athens and Persia, male dominance coincided with the ascendance of what Pericles regarded as the unavoidable brutality of empire. In mythological terms the furies (Erinyes) who haunted Orestes gave way to benevolent Eumenides, and that was the hope: go through tumult to reach peace where peace is the result of uncontested rule. An enduring, safe kingdom is the recurrent promise, renewed in the *pax Romana* as a justification for Rome seizing Carthage and Spain and for moving north into Gaul and the British Isles. Federal unions, too, try to quiet regional turmoil, as the United Nations on a global scale seeks to dampen international squabbles. A few ambitious dictators have expanded that principle into prospects for world rule only to be stopped in the turmoil stage. In the background of centralized rule at its strongest is the bundle of sticks featuring an axe, the *fasces* of Roman rule, in the modern term, fascism. That is what appealed to Xerxes, Alexander, Napoleon, Stalin, Mussolini, and Hitler: uncontested authority even if it requires removing resisting populations. A revised national history with a foggy cloud of sanctimony is a common part of the campaign, ancient or modern. Extremely unpleasant assignments like purges require higher levels of glorification as a reward.

One example will do to suggest the link between imperialism and ascendant male deities, the House of Atreus and the Orestes story. It falls roughly in the Marduk and Yahweh pattern of weakening goddess influence and the patriarchal consolidation of militant authority. In Robert Graves' reading (1960), the stages of the House of Atreus story are two, pre-Hellenic (matriarchal) and Hellenic (patriarchal): "An uneasy balance of power was kept until Athene was reborn from Zeus's head," Graves notes, "and Dionysus, reborn from his thigh, took Hestia's seat at the divine Council." From then on, "male preponderance in any divine debate was assured" (424). The trial of Orestes for the murder of his mother Clytemnestra (in some versions) or for getting her tried for the murder of her husband Agamemnon (in other versions) parallels the power struggles of dynasties and lineage systems. In earlier versions matricide was worse than regicide, Graves finds, but in councils dominated by Apollo and Zeus, Apollo defends Orestes for getting rid of Clytemnestra. Aeschylus and Euripides were "writing religious propaganda" (432) in granting that absolution, "the final triumph of patriarchy" in Athens (432). It looks once again like what was happening on the ground was being sent upward.

One has to speak tentatively even there, however, knowing the importance of Minerva, Juno, Ceres, Diana, Venus, and Sophia throughout the Greco-Roman era and lists of war goddesses that go on for pages. Triumphs of patriarchy in the Near East, the Orient, and Europe do nonetheless indicate militancy featuring war gods. The domination of the Father and Son in Christianity is softened by Mariology, but patriarchal authority shows again in the stern god of the Puritans, the exclusion of women from ecclesiastical offices in Roman Catholicism, and the restriction of the vote to property-owning male heads of household. Male ascendance came built into the great chain of being, the master myth of over a millennium, foreshadowed in the hierarchies of early tribes and city states and fitted to lineage hierarchy. With its passing of estates to first sons, patrilineal descent was typical in most ancient cities and throughout the prolonged era of the chain of being. The *being* part of that is what makes it a supernatural fiction.

Nations with either matriarchal or patriarchal lineage and of whatever size or stage of development were never forced to become militant whatever myths they adopted. The option was always there in households and commonwealths alike to build alliances and share power. The violence of wrathful deities is typical, but so is peaceful behavior from those same figures in better moods. Sophia and Ceres aren't noted for aggression, and the Prince of Peace and his virgin mother are pacifist figures, as was the Franciscan order in following suit. To think in terms of more recent eras, the tendency over the past three centuries has been to change from monarchical rule by a predominantly male lineage to parliamentary legislation. The Democracy Index compiled by the Economist Intelligence Unit estimates the degrees of democracy in 167 countries, and it is higher now than in previous eras. In that count 20 democracies are relatively complete, some 59 are flawed, and 36 hybrid. That leaves 50 authoritarian regimes. What simulated history organizes in all of them is collective energy. It adopts different stories of origin and ancestry, but the rhetorical aim is similar. Idealized icons work better for indoctrination purposes than a realistic history marred by partisan interests. The impressive architecture of national monuments and buildings puts the high status of the union in plain sight.

The founding stories of the ancients were voluminous and varied, and we can assume more were in circulation than got recorded. Near the first major correction in the writing of history in Thucydides, Herodotus puts the beliefs of several surrounding countries on record and illustrates the skepticism with which the propaganda of one nation strikes someone outside it. In the lead-up to the Persian campaign against Greek city states, he assigns Xerxes and Mardonius a typical spectrum of motives for nationalist aggression, which, Herodotus comments, were specious but "made to seem plausible" (418). That is a classical national level feud between ancient rivals. The motives Xerxes advances assign Persia a prominent place in the scheme of things accompanied by scorn for Greece. Citing his ancestry in Cambyses, Cyrus, and Darius, he vows not to fall short of them (416). The ancestral figures are placed in a legendary framework whether or not the details fit actual biographies. Besides avenging Greek insults to his father Darius, Xerxes is working for "the benefit of all... subjects," by which he probably means deeding them Greek land or awarding them shares of transportable plunder, another description of the proposal, stripped of rhetoric, amounting to land piracy. He expects to extend the empire until its "boundaries will be God's own sky, so that the sun will not look down upon any land beyond the boundaries of what is ours" (417), a textbook case of expansionist rhetoric turning an aggressive foray into a heroic mission. Herodotus seems to see it that way, as most victims of an invading power would.

Similarly motivated war councils turn up in Thucydides' Athenian debates, and counterparts again in Tacitus' account of the emperors after Nero–the brutal Otho, the bestial Vitellius, and Vespasian: "The story I now commence is rich in vicissitudes, grim with warfare, torn by civil strife, a tale of horror even during times of peace." Stamp *same* or *similar* on the cover of all imperialist histories.

In retrospect after the toll has come to four emperors slain and three civil wars entangled with foreign wars, Tacitus sums up: "successes in the East, disaster in the West, disturbance in Illyricum, disaffection in Gaul" (1.2). Even where such campaigns proceed as proposed, the resulting empires have chronic problems with rebellions and a continued need for vigilance. That too is typical. To include nations in most eras we could make carbon copies and merely change the names. The US mired in Iraq is a current example except with *freedom* the high power term in that case. In the latter example, fear and paranoia played a role in the initial justification based on stockpiled weapons of mass destruction. When that pretext collapsed, freedom covered the next one, putting it in terms of a necessary regime change. Oil fields and religion were not part of the press releases except insofar as the impetus of 9/11 carried over. Saudi pilots based in Afghanistan were responsible, and so invade Iraq, which lacked terrorists–so ran the argument. Without the term *freedom* the non sequitur would have been even more obvious than it was. Under the on-going exploitation of 9/11, the rhetoric didn't need to be convincing.

Continue with Xerxes. Sycophants are usually on hand to undermine sage advice should reason not be overwhelmed among those being recruited. Herodotus has Mardonius fill that role: "Of all Persians who have ever lived… and of all who are yet to be born, you, my lord, are the greatest," he says, or so Herodotus has him say: "Every word you have spoken is true and excellent" (417). Equivalents to that, too, aren't rare in councils of war if not such patently insincere ones. Surely, the reader thinks, someone in a Greek historian will step forward to spell out the costs and correct the demonizing of the Ionians. In this case that role falls to Xerxes' wise uncle Artabanus, who has previously advised Darius not to invade Scythia in a campaign that ended almost as disastrously as this one will. Waging an unnecessary war has no redeeming value in Artabanus' opinion. A field of equine and human carcasses rotting in the sun argues forcefully against it.

Anyone immune to the fervor of the moment could foresee what Artabanus foresees. Multiplying the motives and rhetorical devices by those of other imperialist powers and dynasties since about the 4th millennium produces an encyclopedia of parallels. *Wikipedia* lists some 175 plus a few unofficial ones. The proper names of people and deities change, but the patterns are recurrent. In the 2003 incident, the Republican caucus in congress played the role of Mardonius, and the Democrats by and large succumbed. In their defense, the combined momentum of patriotism, the 9/11 attack however irrelevant to the case, and a history of the 'greatest nation in the world' complex pushed in that direction. That and the built-in shame component of groupthink were enough to kill and disable people in six figures and add that much more to the unsettling of the region, the US involvement in it, and the recruitment of additional Islamic cult fanatics.

Counter Movements and Setbacks

Setbacks in expansion and imperialism are nothing new, and of course they give the lie to the buildup campaign and its rhetoric. Deflation often follows inflation as devolution does evolution in cultural matters. In the twelfth century (BCE) the ravaging of the eastern Mediterranean by marauding Sea People triggered what anthropologists sometimes regard as an early dark age that saw cities being abandoned and regimes broken. Regression can become nearly as prominent as the advance of civilization. Past retrenchments have been due to invasions, resource expenditure, environmental degradation, hegemony battles that leave both sides weakened, and other matters independent of mythic pasts and promised futures.

That isn't to say that every aspect of glorifying the past is counter productive. Illusions can in fact be highly motivating and the progress they encourage can be quite real in architecture, engineering, literacy, the arts, and mechanization. The story of North America over the past four centuries is a prime example. Civilization has benefitted most segments of the continent's population including those with a relatively recent hunter/gatherer ancestry. The cost has been the overrunning of what exists, genocide at times, and confining hunter/gatherers to reservations. Places with longer civilized histories have had more relapses than the new world has had with its abundant resources. Whenever national fortunes collapse despite being thought to be under divine guidance and protection, the incompatibility between actual history and the sponsoring power has basically only three directions it can take: disillusionment, justification of the misfortune on moral grounds, and postponement in finding the reason for the setback. The Hebrew prophets chose the second, the moral ground, and made national misfortunes the responsibility of those not fully committed to the Yahweh cult. The reliance on mystery handles apparent contradictions by postponing history's making sense until the close. A fourth alternative not much used is that of Malthus, making hardship and suffering purposeful as a way to strengthen character.

Extended periods are seldom free of setbacks. Take the domination of the church and the myths of feudalism, which went a thousand years without a complete overhaul. What was required to keep the Holy Roman Empire intact was ruthless suppression, the Inquisition, and a lords-and-serfs economy that rendered rebellions ineffective. The era was dominated by variants of an alternative cosmos with little resemblance to the real one. In England a mythic Arthurian past in Spenser led to the Faerie Queene (Queen Elizabeth) without much intervening history. Spenser fills out a chart of nation-serving virtues that amalgamates national (Arthurian), philosophic (Aristotlean), and Christian components and confirms the place of the nobility under the monarchy. A transfer of empire from Rome to England, in competition primarily with France and Spain, is part of national myth.

As we've seen in Haklyut, an evangelical strain made its contribution, but the otherworldly goal of a transplanted new

Jerusalem conflicted with it and put a twist in Spenser's Red Cross Knight's assignment to the queen's service. His pilgrimage parallels the world-rejection of Everyman's. Milton wasn't alone half a century later in surrendering empire entirely to industrious Satanists. Medieval allegory had done so as did Milton's contemporary John Bunyan in *Pilgrim's Progress*. The landscapes of spiritual pilgrimages put allegorical places like the Slough of Despond in place of territory that had to be conquered before it could be inhabited. In that Puritan movement lay the seeds of rebellion that together with the Reformation wars in Europe succeeded displaced both the Roman empire and feudalism. Preempting rather than denying a war god supported much of the revolution until the replacement of the former myth-based rhetoric with the secular *liberty, equality, and fraternity* and ultimately in Marxist ideology with an absolutist rhetoric of its own translated downward from an everlasting city where the last shall be first to "from each according to his ability, to each according to his need."

Fact Checking What the Texts Say

Finding out the truth behind simulated history has several branches, but it has taken a common form over the past century in archaeology's comparing of unearthed artifacts with written records where they exist. Only a few early accounts of commonwealth narratives resemble what excavations find to have been the case. The principle searches have been for evidence of Homer's Troy and Hebrew conquests. Exceptional in that regard are Lucretius' synoptic version of the rise of states in *De rerum natura* and Thucydides' in *The Peloponnesian War*. Medieval, Renaissance, 17[th,] and 18th century commentaries are still inclined to fit national narratives into the biblical framework in which 10 tribes of Israel were lost to history and are expected to turn up somewhere in the new world explorations and colonizing.

In that case only DNA evidence would confirm or deny the history since it would have left no trail of artifacts. Myths of decline from gold to iron ages reverse the progress that Thucydides and Lucretius portray, and these too are to some extent evident in excavations. That a golden age actually existed in pre history is one of the recurrent myths used to frame critiques of decayed nations in the manner of Rousseau and Engels. A theory of general decline also governs Hebrew, Christian, and Islamic mythology where the crucial part of the myth is that an original fall has infected life of all kinds. Recovery from it requires redemption.

An iron age decline was the prevailing view in the Near East and West for over two millennia, the most familiar version in the Hebrew anthology with added elements from other myths. Egyptologists and Assyriologists began tabling parallel texts over a century ago. Sa-Moon Kang (1989) finds pre biblical sources not only in Mesopotamia and Egypt but in Anatolia and Canaan. Even so it was predominantly the redacted Hebrew history that European colonialism carried around the globe, with the creation chapters supplemented at times with Greek cosmology as in the creation books of *Paradise Lost*. In making strategic use of biblical texts, the colonies imposed that truncated natural history on places and people whose racial histories were far older than the chronicle specifies, as of course everyone's was in natural history. Anything that runs as counter to evidence as myths of origin and decline must have a rhetorical purpose, and coordinating groupthink is the main one that comes to mind.

Richard Haklyut's Elizabethan naval chronicles (1589-1600), to return to an English example, assumed the reigning world history that European countries had in common. It too put natives of the New World and the people of Asia in the biblical framework. The voyages apart from Haklyut's account were the first real opportunity European sea goers had to correct misconceptions of world geography and to glimpse the course humanity had actually taken in migrating around the world from Africa. Haklyut instead wrote a hymn to

English exploration and its evangelical mission. Underlying European expeditions was a desire for trade and wealth, as mercantile interests infiltrated and began to replace royalty-granted patents with wealth funneling more into a rising middle class and less into the elite of the monarchies.

Professed beliefs transported in several European variants needn't have been sincere to play that role. Hume's comments on Cromwell, to take a land-locked English example, finds Cromwell using religion, in Andrew Sable's words, in "a classic princely mode, leaving the degree of his zeal unclear and seeming more religious to some than others" (Krause and McGrail, 250). *Classic* draws on what historians have found elsewhere and reflects Machiavelli's views in *The Discourses*. Where religion exists, the latter maintains, "it is easy to teach men to use arms, but where there are arms, but no religion, it is [only] with difficulty that it can be introduced" (1.11, 140). Religion adds its own rewards and indexes of shame to the rewards and punishments of the civil order. Edward Gibbon (1788, 2000) recites that principle cynically and wittily in remarking of the Roman world, "The various modes of worship… were all considered by the people, as equally true; by the philosopher, as equally false; and by the magistrate, as equally useful" (35). Without finding its way into the records, that had likely been a common attitude of the streetwise in many parts of the globe. That is again not something that would leave a written record or artifacts but the vigorous condemnations tell us it existed.

The Hebraic myth/history hybrid has the most often analyzed mix of invented and historical factors, which altered its original nationalist focus when Christian and Muslim evangelism went international. Supplied with previously written materials, the Deuteronomist(s) compiled what amounts to an epic of state. Israel Finklestein and Neil Asher Silberman (2001) suggest strategic reasons for retrofitting the past to the times by whatever redaction was necessary. The Bible's historical saga extending "from Abraham's encounter with God and his journey to Canaan, to Moses' deliverance of the children of Israel from bondage, to the rise and fall of the kingdoms of Israel and Judah" wasn't a "miraculous revelation, but a brilliant product of the human imagination." It was also strategically rhetorical as an instrument of contemporary rule. That constituted a new 7th century movement of "unparalleled literary and spiritual genius. It was an epic saga woven together from an astonishingly rich collection of historical writings, memories, legends, folk tales, anecdotes, royal propaganda, prophecy, and ancient poetry" (1-2).

Whoever the initial Deuteronomist(s) might have been, the division between south and north kingdoms and the territorial ambition of Judah set the current conditions of the redactions: "Much of what is commonly taken for granted as accurate history–the stories of the patriarch, the Exodus, the conquest of Canaan, and even the saga of the glorious united monarchy of David and Solomon– are, rather, the creative expressions of a powerful religious reform movement that flourished in the kingdom of Judah in the Late Iron Age" (23). The mixture of narrative modes compresses simulated world history into that of a single nation of comparatively recent founding, recent by comparison to city and territorial states of Sumer, Babylon, Persia, and Egypt.

Along with textual critics, archaeologists have concluded that the Canaan state was breaking down of its own accord for internal and external reasons and was never invaded by united Hebrew tribes. The militant part of the myth in that perspective was an invention designed to reunite a divided kingdom. In citing excavations of 23 of the targeted cities, Lawrence Stager (1998) finds the exodus story implausible (134) on the grounds that it "fails to meet minimal standards of criterion 1," namely, that Israelite artifacts be found immediately atop those of indigenous culture. His reasons are basically those that Carol Meyers (1988), too, advances, namely the prominence of heroes and their myths in national cultures, especially in early days when they are insecure. By the time of the Deuteronomist(s), that use of an enhanced national legacy had been around for centuries. The covenant with the maker of the universe forced Christian and Muslim sects to set the original chosen nation aside in order to claim that source for themselves. They didn't hesitate to do so in battles for empire. Converting the Jews to their way of thinking was an alternative to a conquest of the holy land and a vital part of the evangelical mission, though in the crusades and at other times, taking the holy land, too, was part of the campaign. Once Cromwell's saints succeeded in displacing king and court they were to set to work on that mission, so the millenarians among them proposed. Assuming control in England was only part of the project.

Because Milton held the Eden story to be true in essentials, that becomes a prolonged historical problem. That the redemptive sacrifice of the Son didn't come sooner is due to history having gone centuries without producing a Messiah. History preempts the logic of the story telling, which provides no reason for the pre and post Messiah delays extending for centuries. After the Messiah's incarnation, more centuries pass without an outcome. Milton's epics can come to a conclusion only by foreseeing the end of time, something poets, common people, and theologians alike did in the wake of *The Apocalypse of John*. Predicting the timing proved to be problematic. A good many tried and failed. Whenever it came, the end of the cosmic story was to bring a definitive sorting out. The defeat of Satan in *Paradise Regained* is just that definitive victory, and yet history again goes on as if nothing had changed.

A premium on free will didn't serve Milton or his sources well as a reason for the delays, since it can't hold omnipotence back and it could as well be exercised in ways other than rebellion and produce results other than mortality and universal suffering. Limits imposed on it in a restored paradise could presumably have been in place at the start with no more cramping of self fulfillment than will be the case in a hereafter. Moreover, an infallible power that demands obedience would presumably have created only obedient creatures to begin with. Those who will eventually be saved must be considered worthy of that end, but then they would also have been worthy of not suffering in the first place.

That is only one of the dilemmas that come of putting national histories in the context of myths. It has counterparts elsewhere as well. In *The Aeneid* much of the second book is devoted to retelling Homer's Trojan horse episode with divine intrusions of a different magnitude seeing to the destined history of Rome. On the face of it a wooden horse loaded with enough warriors to take Troy shouldn't deceive anyone, but the Greek Sinon is a convincing actor, and what he claims receives timely confirmation from the gods. As the one whom the Greeks have selected to sacrifice he has discovered that "crookedness/ And cruelty" are afoot for him in the Greek camp (2.169-171). That provides a plausible reason for him to cross over to the Trojans and sets up his claim that the wooden horse is a votive offering. As such it should be moved inside the gates (2.345). Marvels staged by Minerva discredit the two Trojans determined not to let it in, Laocoon and Cassandra. Minerva punishes Laocoon by raising two serpents from the sea that devour him and his sons. She, too, clearly wants Troy to fall so that the seat of empire can migrate to Rome and move onward from there to further extensions of the Augustan empire. The entire episode–the actors, the signs, and the fall of Troy–hangs on divine intrusions on behalf of a destiny already confirmed by Roman achievements in Virgil's day. Augustus' patronage of poets, designed to enhance his authority and his place in literature, shows in the lives of Horace and Livy as well as in *The Aeneid*. Whatever fables the poets concoct, the goal for Augustus is to further the empire and solidify his place in history. It is up to the poets and the oracles to provide the appropriate simulated history. The gods are again among the legend amplifiers, Augustan megaphones.

That is a common pattern of national level illusions designed to erect an umbrella high myth narrative over citizens otherwise likely to be subdivided into ranked orders, regional groups, and other partisan interests. The common heritage provided for them in the national myths connects the nation in a direct line not only to its celebrated heroes but to no less than the overseeing of the universe.

Mankind is once again placed at its center, and once again moral philosophy preempts natural history's visible universe to make it fit human terminology.

Historiography vs Myth/History Hybrids

I want to return to examples in Thucydides that, by running counter to history rewritten for nationalist purposes among the ancients, apply standards more common to historians who rely on documentary evidence. In method his account of the Peloponnesian wars could stand beside Edward Gibbon and others of the past two centuries. *The Peloponnesian Wars* not only includes an encompassing framework of early Greek days at the beginning of agricultural settlements but eliminates divine interventions except insofar as influential people who believe in them have to be cited. Hegemonic contention is at the heart of the Sparta/Athens conflict. Athenian imperialism puts local autonomy in peril through much of the Mediterranean, as Persian militancy had as well. After the Sicilian debacle the Athenians have no prophets to shift the blame for defeat to the people as Hebrew prophets often did. Rather the reverse: the demos are "furious at the oracle-mongers, seers, and anyone whose divinations had made them hope that they would capture Sicily" (411). Where both sides of a war appeal to oracles, they will fail half the time.

The historian himself is reporting and offering commentary, not serving as a moralist. He echoes Homer and the nemesis pattern of Greek tragedy only to upend them. Once the Athenian defeat in Sicily is assured, other city states that expected to be attacked had Athens prevailed go on offence. In doing so they aren't fulfilling destiny in Thucydides as they might have been in tragedy or epic but capitalizing on tactical errors. The decisions are timely and circumstantial, but in the framework established in the introduction they are also a phase of territorial state formation. Moving from city states to nations inevitably brings hegemony struggles of the kind that Sparta, Athens, and their allies illustrate. Military intelligence, weaponry, logistics, and numbers in Thucydides determine battlefield outcomes. Self interest determines policies and is defined in terms of individual ambition and city state management.

The Melian dialogues are a case in point in policy debate using intimidation or what would now be called leverage. Despite what Nietzsche (2000) misreads as a discussion among equals where "there is no clearly recognizable predominance and a fight would mean inconclusive mutual damage" (148), the Athenian delegates have a decided advantage. To head off disaster the Melians must either get help from allies or capitulate, and for various reasons help from allies won't be forthcoming. As a consequence of their resistance, Melian adult males are killed and the rest of the population reduced to destitution. Their dilemma initially is basically that of any group threatened by a dominant power whether the enemies of Sargon, Athenians overwhelmed by Persians, or Polish refugees executed by Stalinists. The Melian suicidal choice has sometimes been judged harshly, but the inhumanity of the Athenians bears the chief responsibility, and in the heat of the moment courage can't always be distinguished from recklessness, one version of which is the self mutilation of a scorched earth retreat. When the underdog wins against odds, the cry "give me liberty or give me death" looks less foolhardy. Hezekiah and the people of Jerusalem are usually praised for fighting on against the odds even though it takes a miracle to save them (*2 Kings* 19.35). The Jews of Germany are sometimes blamed for going too quietly when resistance would have meant expedited execution and required organization and arms that were impossible under the circumstances. An alliance of Greek states at Thermopylae made one of the most famous last stands in history, suicidal though that, too, was and completely different from Custer's last stand, foolish from start to finish.

No one in the dialogues examines in any detail what empire building might entail if concepts of destiny and necessity were stripped aside and only what will benefit allied parties were considered. It doesn't occur to the Athenians that they might rule in a less abusive way or that shared governance might avert retaliation. The choice lies with them and depends on whether they want area accord or empire at any cost. A nation composed of scattered islands and cities is a conceivable alternative, a league like the former Spartan League working toward a centrally governed group of semi sovereign cities. That is a pattern we recognize from numerous versions of a governing center and outlying districts. It can be enlarged to a continent or shrink to a few districts. In retrospect from the disaster at Syracuse, the invasion of Melos was a warning sign. Forcing a union for the wrong reasons raises resistance, much as colonies dominated by colonial powers felt too restricted by their satellite status to be satisfied with it.

Thucydides appears to see it that way, though in Simon Hornblower's view (1987) he only half realizes what greed and arrogance bring: "What he failed to see was that the 'truest cause' of the war, which he correctly diagnosed, and the 'truest cause' of the Athenian defeat, which he incorrectly saw as the greed of a particular generation of Athenians and politicians, were identical: the empire was built on greed" (176). Plunder often does figure in subordinating formerly sovereign powers. Statecraft is in short supply, greed abundant. It could be that a plan for consolidating a federal state would work without conquest. Remove the pillage factor and see. The spectrum of precedents in consolidating smaller into larger states needed to be broadened. Conquest wasn't the exclusive way to unite dispersed settlements. North America doesn't present an ideal model in that regard because territory that could be apportioned peacefully to provinces and states and to immigrants from Europe had first to be taken by force.

Greed is accompanied by indifference toward inflicted suffering, another foreign policy mistake when it comes to long term relations. A population beaten down doesn't forget or forgive. The most despised segment within a former sovereign state will likely be its quislings. That greed is common suggests that nothing very special by way of circumstances is needed to trigger it. It is the outgrowth of basic organism self management. Where power and profit team up, motive disguise and upgrading are the rule. Godly amplification turns on. Nothing more is needed. Those who have the military advantage are only too likely to put rhetoric to good use whether or not it applies any particular rationalization. Racism and nationalism are available as well as theology. Sending an armada off to Sicily isn't all that different from earlier Athenian aggression or from Persian invasions within memory, rendered by Herodotus from the Greek point of view. To other pragmatic concerns add long supply lines and unexpectedly stiff Sicilian resistance.

Both Pericles and the Lacedaemonian Archidamus underestimate the enemy at the beginning of the war, and the Athenians underestimate the Sicilians. That optimistic misjudgment isn't a necessary part of a conquest, but it often is. In both the internal politics of the Athenians and in the diplomatic front they present to the Melians, Megarians, Cocyreans, Spartans, and others, the overweening attitude stands out more than one might expect in an era of such extraordinary minds and teachers as Socrates, Plato, Aristophanes, Xenophon, the tragedians, and Aristophanes. It indicates again that neither a supreme fiction nor an ideology is necessary to generate an infectious partisan spirit. What isn't allowed a voice is a concept of governance based on alliances and delegated authority, with fair trade in the mercantile class replacing plunder in the military class.

So modern seeming is Thucydides' history that what he says of Athens can be applied to similar cases all the way to the present. Walter Karp in "The Two Thousand Years' War" (Blanco 1998, first published in *Harper's Magazine*, 1981) suggests several such applications in the modern era. Bickering similar to that of Athens and Sparta escalated a minor incident into War I. Abuses of power brought on War II. Over confidence based on rhetoric pushed the Truman/McArthur army northward in Korea. Over confidence caused Kennedy, his advisors, and Linden Johnson to launch the Bay of Pigs and Vietnam fiascos. Had Karp written later, he could have added the Bush, Cheney, Rumsfeld, Rice attack on an Iraq that was overmatched militarily but certain to become a drain on resources for years to come. The toll by December 2011 was calculated at over a hundred thousand Iraqi civilian casualties, numerous cases of post traumatic stress syndrome, and nearly a trillion dollars in costs extracted from future taxpayers. Concluding that it wasn't worth it has followed many a conflict. Unfortunately, international tribunals are rarely influential enough to administer justice and discourage future transgressions. War criminals who remain in power generally escape all but verbal reprimands. Prewar rhetoric is usually followed by postwar grief. When the amplifiers are turned off and the air released from the icons, what is left resembles the collapsed plastic Santas and reindeer on post Christmas lawns. Disillusionment is an almost inevitable conclusion to illusions. The invention of an anagogic level that fulfills promises elsewhere was a brilliant stroke in that regard. An anticipated apocalyptic moment can be shown to have been mistaken but not what it was to have produced can be postponed to another moment and be shrouded in mystery or administered individually out of sight.

 The Athenian and Spartan fervor had no serious doctrinal element, but other illusions and self deception took their place. The overall impetus came from imperialism, an enlarged version of tribal partisanship. In its presence advocating peace comes to seem soft. Though sometimes strong in judgment, Thucycides' editorializing is level headed in lamenting the Athenians calling irrational recklessness courageous commitment; in calling anger true manhood; hesitation, cowardice; moderation, lack of manhood; and circumspection, inaction. The man of violent temper seems credible while those who oppose him are suspect (3.82-83). Concerning the council of Athenians and Melians, Thucydides reduces the gruesome results to a single sentence: "they killed all the grown men they captured, enslaved the children and women, and settled the place themselves by sending out five hundred colonists" (5.116), a pattern repeated numerous times in the prolonged transition from nomadic tribes to city-dominated civilizations. In this case responsible historiography retrieves the details that prewar rhetoric omitted. Given the War I change in battlefield reporting, the use of cameras and better record keeping, we have gotten used to a different kind of documentation, but in Thucydides' time aftermath reporting was more likely to resemble that of the Moab Stone, the Tiglath-pileser inscribed boast, and Hebrew myth.

Using cover stories may be rational in the sense of expedient and based on statecraft, but rational and principled aren't the same thing. Shrewd and unprincipled are warped images of them, and they combine again in principled irrationality and unprincipled street smarts. Any of these combinations can lead to conflict, with those in the combat corp as unlikely to resist successfully as a remuda on a line. What indoctrination, maneuvering for power, ideology, and imperialism have in common is glorified group history combined with shame for dissent. Were more factors and a longer segment of history brought into view, the judgments would be less partisan and irrational recklessness less confused with courageous commitment. From the naturalist perspective, knowing what the natural history continuum contains is to epidemic beliefs what inoculation is to communicable diseases. Once an idea epidemic gets underway it is difficult to avoid Thucydides' inventory of perversions. Irrational anger will indeed seem true manhood and hesitation cowardice. Bigotry will seem patriotic and tolerance weakness. Conversions of one thing into another are the essence of militant rhetoric, human dementia into divine favor, ideological rigidity into high principle, foes into demons, the normal run of differences into furor. Escalation and inflation are the principle and hyperbole the rhetorical means. Inflation to utmost nomenclature carries that to the maximum.

Chapter Ten

Groupthink on Steroids

Public Exposure

The obstacles to national unity are numerous and never merely negligible. Nations are crazed with fault lines based on region, beliefs, ideology, age, sex, and class, plus the usual personal attractions and repulsions. That tempts them to group people for specific purposes by whatever means they can, coercion, ceremonies, enhanced history, shame, commendation, glittering generalities. Like all categorical titles, a nation's is a verbal contrivance that ignores the divisions within. The people overlap and collaborate in matters of mutual concern but are many things else. The divisions and links among them are further complicated by the means of communication, none of which makes a full display of a given segment. National unity now faces obstacles that have arrived with postmodernism, increasing numbers of translations among languages and distance-defying transmission of ideas and information by electronic means. These join people in ever-more tenuous webs within and beyond nations. In modern collectives those equipped with videos, newspapers, books, and software can cultivate and break off links by the dozens. Even the miscellany of an ecological wild area has nothing quite comparable to that message driven volatility.

Considering individuals single units is itself a simplification. The brain's internal filaments are staggering in quantified terms, 100 billion neurons, supporting glia cells linked into a thousand trillion synapses with enough mileage in fibers and filaments to stretch several times around the world. With such networks inside us we have no trouble being inscrutable. If as Heraclitus said we never cross the same river twice, the same person never crosses twice either. The components of a normal brain and its connections to reality and to other brains are beyond numbering. That doesn't prevent public displays of unity in whatever congregation is relevant to the moment in what we can call the stadium effect. The sight of a hundred thousand assembled to hear a firebrand speech can be terrifying when the speaker is a Hitler and many in the audience are dedicated members of a war machine. Groupthink has a way of simplifying one mind in coordination with others likewise simplified. The unanimity can be based almost entirely on illusions as that of the Third Reich was, which is what involves myths of the commonwealth. It is at the national level that governments are strongest and most compelling. They pass laws, raise armies, and deploy them on the basis of their favorite stories.

That the public affairs of nations are conducted behind facades isn't itself among the well-kept secrets. We assume false fronts in politics and other public representations and learn to allow for them. In parliamentary democracies with neither an orthodox doctrine nor a rigid ideology words like *liberty, equality* and *fraternity* take the place of regal pomp and circumstance as a public front. Though stated as if absolutes they are enacted as conditioned practices, with *fraternity* less tangible, for instance, *equality* referring usually to legalities, and *liberty* meaning at a minimum not subject to search, seizure, and arbitrary imprisonment. Like everything living and not living on whatever scale, a nation too must have a binding force. That it is even truer of literary coherence than of history underlies the application of storytelling logic to national histories, the key to the reconstruction of a national past in its retelling in literary form. The structure Shakespeare imposes on the history of Henry V, for instance, serves Elizabethan England as its heroic heritage. Good dramatization makes for plausible national history taking shape under an icon. Nothing like Falstaff accompanied the rise of the real Henry V, only a Lollard named Sir John Oldcastle, but Shakespeare needed a foil for Prince Hal and set Falstaff up to be rejected when the new king can no longer be a playmate to folly, which the real prince once was. The focus he brings to the nation points it toward France and Agincourt. It also points toward the reign of the current monarch. Intricate and detailed historical novels and epics make similar uses of chronicle events as springboards, contributions to a national identity. It is a brief journey from group allegiance based on valid information to edited propaganda. That is sometimes obvious from the outside, as a late 16th century French reader of Shakespeare's *Henry V* would see it differently than contemporary London audiences did. The fictional elements plus groupthink enthusiasm are power boosters. Thomas Keneally in *Searching for Schindler* (2007) recounts that the Nazi execution of Jews was labeled Special Treatment. Screening potential victims was Health Action. Personal assignments to that SS task were to Special Duty Squads (111-112). Those thus assigned and doing the assigning knew or at least suspected what lay behind the euphemisms. After-the-fact history denial joins science denial as a mental device to substitute 'alternative truth' for what sane and sensible people see as real fact.

Public discourse and election campaigns in democracies use euphuisms similar to those of propaganda ministries and spin doctors. Without strong declarations from the podium and answering applause, a candidate and a party platform have little chance of being adopted. We've no reason to believe that modern states differ greatly from older ones in that respect, but because they involve larger populations their propaganda has to be more proficient and far reaching. Transferring some of the togetherness of clans to larger collectives requires not only communication but either sacrifices of local and personal interests or hope of personal gain. As Engels points out citing Alfred V. Espinas, "before a herd can be formed, family ties must be loosened" (63). Not only family ties but individualism before them and tribal ties after them. *Brotherhood, sisterhood, fatherland,* and *motherland* transfer familial feeling to patriotism, sisterhoods of nuns, and brotherhoods of monks. Where biology and psychology have created an internal wiring for intimate relations, social orders put the details in writing. The biological and psychological givens can't be changed, but the rest can be drafted, crossed out, and revised.

It is at the national level as I suggested that myths do their most work, but nations in turn are directed by ideology and doctrine, and these easily cross national boundaries. Among the ancients they were more contained by borders, as Israel had one set of beliefs, Babylon, Egypt, and Athens other sets. In the later 19th and earlier 20th centuries, Marxism and capitalism became the two dominant ideologies of machine and factory production. They became ideologies with the pervasive influence formerly possessed by monarchies and religions. In Marxism class allegiance overruled or joined with geographic and clan ties. Capitalism was tolerant of other beliefs as it was of democracy, oligarchy, fascism, and even monarchy. Its incentive for individuals in parliamentary democracies was a chance to gain wealth and social standing. The incentive in Marxism solidarity and a fair wage. Both had disadvantages and were in competition with non economic priorities. One of the sources of class tension in under-regulated capitalism from the 19th century onward became extravagant extractions of funds by those in finance positioned to siphon them off as they passed through the system. The class fissures among any large number of people become more pronounced where a social order lavishes favors on an elite. In capitalism the emphasis on self and the ownership of stocks increased the money flow between private sources and corporations. Unions formed to keep more of the value in the hands of those who added it to raw materials, including invention, design, and marketing. Rewards nonetheless become unbalanced in cycles on behalf of a sector that doesn't invent, produce, transport, or market, merely shifts funds from one place to another. The ratio of earnings between higher and lower positions in all sectors combined, according to Economic Policy Institute calculations, grew from 24.2 to 1 in the later 1960s to nearly 300 to 1 in the year 2000 with robots added to the mix in modern assembly positions.

The generalizing and sloganeering that help unify diverse populations are clear enough, ancient or modern. What they too obscure are internal group differences and the maze of connections that link individuals, businesses, and interest groups. Alliances and enmities come and go within any national identity. Though the personal computer wasn't the first means of making distant contact it has been one of the more proficient networking devices. Adding immediacy and distance to conjunctions made them quite different from a reader consulting a publication. Where a telegraph or telephone requires a live presence at each end of a connection, transmitting and recording devices tolerate pauses and delays. Residential television dispenses edited information and proliferates illusions and fictions as well as news. It, too, when equipped with a recording device is tolerant of delays and planned timing in the reception. When electronics went small in notebooks, tablets and cell phones, the transportable convenience added to the maze of links of individual hookups. Flexible consulting means multiplied conjunctions. The filaments lengthen and the networks get more complex.

The scattering of connections is partly what social orders counter when they insert dominant presences as gravitational centers. In place of tyrants and their minions using tread-worn myths, modern systems in open social orders use celebrity glamor and hyperbole, but even in dominantly secular cultures they haven't completely abandoned the use of transcendental fictions at the high end of groupthink devices. A national figure campaigning for election normally displays one or more of the older pieties, which across the social order blend into newer ones in palimpsest confusion.

Beehives and Bands of Brothers

To activate instinctive sociability, animal kinds require little more than nudging, chattering, honking, buzzing, mooing, and howling. No such automation works for symbol users splicing connections together, though they too have instinctive chemistry at work. The attachments draw on mammalian bonds millions of years in the making. They survive transmission even in handwritten and printed form despite the lack of voiced presence. Squeezing them into thumb-tapped tweets is not very imaginable, but small amounts can miraculously survive the passage. To simplify the levels at which attachments work: the brain features the cortex, corticolimbic, limbos, and reptilian survivals. In the deeper and older brain, instinctive elements dominate and are more felt than understood or translatable into language. The cerebral cortex specializes in ideas in superstructures erected beyond the reach of simplistic feeling.

Myth studies initially gave little attention to the combinations of older and newer brain segments or to the distinction between to consciousness and the unconscious. Mircea Eliade's "myth of the eternal return" (1954) featured dreamscape heroes who left for strange lands and returned altered. Their bonds with the originating society changed as a result. Scholars such as Lord Raglan and Otto Rank emphasized the psychological underpinning of such archetypal figures. Oswald Spengler's groupings (1926), to which he gave such titles as "Egyptian man," were more national and ethnic and based on geography and culture rather than on subconscious archetypes. How a national difference connects to archetypes and why being an Egyptian is an emotional as well as a territorial matter wasn't Spengler's concern, but nationality can hook articles of belief onto archetypes, as a mythical Moses with Egyptian and Mesopotamian predecessors supported a Hebrew priesthood founded in his name and (along with Abraham) became a national archetype used to support laws and ordinances. Myth thereby worked down into domestic and foreign policy and put an edge on relations with Babylon, Assyria, Tyre, Egypt, Persia, Greece, and Rome. Along with a good many other stories such as the creation, the flood, Noah's arc, and crossing through the parted Red Sea, myths defined a national heritage by means of texts that were anecdotal. Legislative, advisory as books of wisdom, poetic, and historical. That some of the stories are incredible and miraculous made them speak all the more to nationality. No other nation had anything quite like that simulated history overseen by the maker of the universe.

Ancient Israel in that respect was entirely typical in using legends to strengthen the ruling superstructure. A group identity of that kind can stay intact despite tribal differences and even prolonged periods lacking a home territory. In the transition from hunting and gathering to agriculture, other nomadic groups sought comparable home territories and lasting peace. No doubt like Aeneas' Trojans some of them carried household gods with them in articles of belief and idols if not a monotheist power. Sacred ornaments and texts on wheels have to substitute for temples. When an illusion-bound group succeeds in commanding a territory it is likely to build shrines where ceremonial recitations can renew the collective bond. The tangible icons and architecture then give solidity to what otherwise lacks an enclosure for diverse parties to becoming a congregation. An enclosing edifice also shuts out natural history. A large one gives the impression of magnificence, whereas in the open anything human is dwarfed. Inside a cathedral, storied stained glass windows make it easier to imagine a few thousand years a long span and legends actual history. Ancient and modern examples are the same in that regard.

As we've seen, myths are highly transportable and infectious. The colonial era from the 15th to the 20th centuries provides several typical examples of national illusions transported abroad. Haklyut's evangelical nationalism (glanced at earlier) is a case in point, with a Protestant underpinning combining articles of belief purporting to go back to the beginning of the universe, in story form a few generations ago. Along with trade the English have carried abroad, Haklyut believes, "the truth of Christianity, and of the Gospel." That mission has remained intact wherever it has gone: "as in all former ages, they [the English] have bene men full of activity, stirrers abroad, and searchers of the remote parts of the world, so in this most famous and peerlesse government of her most excellent Majesty, her subjects through the special assistance, and blessing of God, in searching the most opposite corners and quarters of the world... have excelled all the nations and people of the earth." What monarchies before this one, his rhetorical question goes, have taken their banners and their faith as far as the Caspian sea, Constantinople, Syria, Aleppo, Babylon, Chili, Peru, the Cape of Good Hope, China, the Philippines? The goal is to use the military, trade, and gospel truth to create a network that combines trade connections and civil administration with Protestantism under a home-based monarchy. Similar evangelical missions carried Catholics and Muslims into South America, Asia, and Africa. The national banners varied but the imperialist missions were similar.

Immigrants and colonizers were sure to encounter resistance in undertaking such a mission. Whereas archetypal journeys go out and return, colonial ventures go out, add satellites and provinces, and renew the age-old center-perimeter tension. That too resembles the Israelite one with its perimeter encounters with what were considered abominations, not unique to Jerusalem but typical of other urban centers of empire, Babylon, Athens, Rome. In the colonial era where a pre-existing civilization was as advanced as India's, justifying the invasion and conquest had to seek out or invent flaws in the native population. In nations that have slipped in influence the bigotry seems to strengthen, perhaps because it has to assert what once came more naturally. At the other extreme, to back off home customs and adopt those of a foreign country can bring an identity crisis. When homeland values crumble, the explorer, as in Joseph Conrad's parable

"Heart of Darkness," can go native. Back home that is frowned on as a calamity rather than a necessary or sensible adjustment to a prevailing way of life. Kurtz turns the archetypal journey inside out and rather than carrying enlightenment forth enters a dark abyss and never recovers. The Hebrew exodus story and Babylonian exile are the opposite in staunchly maintaining group identity despite lacking a home territory. The projected eternal New Jerusalem of Christianity did something similar for colonial evangelism except for transferring the eventual homeland off the planet. It had learned early on to do that when it too lacked a homeland and a ruling regime. It acquired the latter with Constantine and so could advance into alien territory. The final home was where the ultimate congregation was to gather, an interracial and non national gathering. Among the strife-raising things it would leave behind were territoriality, the need to produce and gather food and find shelter. Biology in general.

Despite the assumptions that sailed with the colonial parties, much that was exotic to Europeans flowed in reverse. That and perimeter rebellions were nothing new. Benedict Anderson in *Imagined Communities* (2006) writes of the displacement of the earlier dynastic languages, Greek and Latin, in plebeian uprisings of the vernacular. While transporting Greek language and culture abroad, Alexander was in turn absorbing foreign influences. Rome conquered Greece only to adopt many things Greek. The Roman empire over centuries fostered exchanges on several fronts, African, Near Eastern, Germanic, Gallic, Anglo Saxon, Celtic, Hispanic. Cultural exchange on a lighter note is the subject of Melvyn Bragg's *The Adventure of English* (2003), which uses language as an index of two-way international flow. British English was altered into American, Canadian, Australian, New Zealand, and South African English and became barely recognizable in Singapore's Singlish and China's Chinglish. In North America, pigeon dialects worked English phrases into Indian culture and collected tribal terms in exchange. People in the field *will* talk across barriers as lovers will through chinks in the wall. Ethnic groups–Korean, Vietnamese, Japanese,

Chinese, German, French, Irish, and so on—tend to maintain ethnic distinctions as immigrants but also end up speaking the new language and attending the same classes in math, grammar, civics, and history.

What molds different population into a nation where myths are incidental is long term proximity, speech, manner, and education. Preconceptions and network connections lodged in the mind in the former setting fade. Even where an us/them polarity remains divisive, a momentary bond with 'them' allow a pause in the hostilities and possibly a spread of fellow feeling. Nothing is inherently good or bad about that except to racial and ethnic purists. Cultural practices and habits are no more than cultural practices and habits until they become rigidly prescriptive. One of the most famous water hole truces in history occurred in a western front 1914 cease fire in which German and British troops exchanged seasonal greetings and gifts and played games in no man's land. Both sides tapped into a fellow feeling nurtured in the seasonal ceremonies of Christmas at home. Some substantial after effects came of that particular truce, including an exchange of prisoners and commemorations in films and in songs like John McCutcheon's ballad "Christmas in the Trenches."

Truces and temporary sociability don't make it significantly easier for an outcast or a transplant to change environments. How difficult that can be, especially in older migrants, is the subject of a good deal of immigrant literature. One recent example, the movie *Brooklyn*, features a young Irish immigrant achieving reasonable comfort in New York. Not every expatriate is that fortunate any more than a receiving nation is comfortable with its immigrants. Assimilation works one mind at a time making multiple contacts. A legal system and impartial justice have to be in place to allow time for assimilation. Under those conditions high myth groupthink is less an issue. Influences pass back and forth a little at a time as they do in the acquiring of facts and theories under empiricism in the classroom. Normal daily protocol and manners replace congregational ritual. Music from a radio in the kitchen is the transmission of the day rather than an anthem in the stadium or cathedral.

Modes of Production

A good many factors distinguish one way of life from another, but the most basic is how food and shelter are procured. When residential settlements were first established, they were vulnerable to incursions from those whose nomadic life required unoccupied territory. The ways of life were incompatible for mode of production reasons. In Thucydides' archaeological introduction to the Peloponnesian war, pre-polis Greece is a nightmare of lawlessness begging for an overall authority capable of defending against incursions. That is in accord with what is known of dark age roving parties invading urban strongholds. Where settlements had enough wealth to tempt pillaging, pressure from marauding bands helped motivate broader area alliances to defend against them. Outlaws, Indians, and cavalry on the American frontier repeated the skirmish pattern, in that case initiated by the migration of farmers, cattlemen, and town builders accustomed to settlement life into areas open to hunting by Native Americans. In what in Thucydides were to become the city states of Greece, coalitions of scattered populations made agriculture possible only by combining their strength in defensive networks. Negotiated confederations could conceivably have continued into a broadly territorial state, but because of rivalry among the city states that didn't happen until Philip of Macedon in the fourth century. The era of city state turmoil is Thucydides' subject, featuring Athens and Sparta but dragging in others as well from Persia to Sicily. City state turmoil in a region prefigures national turmoil in larger areas with similar group dynamics at work. In a sense tribal feuds, Montague and Capulet disputes, and world wars are related group-fueled turmoil. Filtering out evidence and acting on stereotypes underlies the commitment of the self to a collective identity.

So prominent are chauvinism, imperialism, and the plundering that goes with them they might seem instinctive, and it is true that something similar to the Greek situation had been the case for centuries in Mesopotamia. The tilled areas there too were exposed to incursions from the perimeter. Once city states were established, imperialism led them into organized warfare. That imperialism and plundering aren't universal, however, indicates that they are avoidable, merely easily triggered. Among the Greek city states, the question of where to build walls was still being debated in Themistocles' era and to some extent in Thucydides' at the end of the 5th century. The most feared incursions by Thucydides' time, as in ancient Mesopotamia, came from rival states rather than from pirates and Persians. Again, however, territorial states and federations could be founded peacefully. For federations of national size, negotiated treaties and vassal agreements were early on among the alternatives to population removal and enslavement. As Carol Meyers summarizes, thinking more positively in terms of mutual defense, "The early Iron Age village communities would not have survived without the protective and integrative function afforded by the emergent state.". Every large republic illustrates the results of the pressure to merge, not always under compulsion or conquest. Early states kept no durable records, but unless human nature has changed radically a Hobbesian total surrender of individual prerogatives to absolutist rule would have come only from conquest. The take no prisoners genocidal policy was prominent enough and lasted long enough that war historians are confident of it.

Oral and Written Regimentation

We have come to associate authority bolstered by legend and myth with texts, but oral dissemination accomplishes much of the indoctrination and establishing of the laws and ordinances that keep incivility to a minimum. Disputes can divide even bands of brothers. Given the already advanced stage of empires by the time of the first inscriptions, we can assume that oral indoctrination predated written laws by several millennia. Though those on the books of any large polity are almost too much to imagine going unwritten, among areas that minimized or did without texts Don Brown (1988) lists ancient India, China, Southeast Asia, and some of the bronze age Near East. In caste societies, patrons "as hereditary monopolists of power and authority" could "say anything about themselves without fear of contradiction, except from their peers, who, of course, benefit equally from any claim to exalted status for themselves" (325). That was the indoctrination side of oral dissemination. Meanwhile street and household talk would no doubt have continued their undramatic work of conformity and resistance to it.

 Even where laws and exhortations are written or delivered at length people still receive them largely by indirection. All but the legally trained remain innocent of law books. For the rules and regulations of traffic, signs in a few large print words in green, yellow, and red take the place of volumes of fine print in black and white. Electronic equivalents in legalese capitalize a few things clients are expected to read before they accept the terms. The expectation is that almost no one will read all the agreements pertinent to a binding contract. Oaths of allegiance and of office are also abbreviated versions of the laws on the books. The enforced obedience and conformity of dictators and ideological regimes require a cadre to do the enforcing. Fascist and communist states are the notorious examples, but dictatorships are never scarce among the world's 190 plus nations. In oracular recruitment where logic is missing and articles of belief contradict natural history, it is left to dramatic delivery to drum up emotion. Objectivity drops and self or group centered benefits rise, one of them escaping the consequences of disbelief. To enter an idealized city of the imagination, one need do nothing to build it, and no laws are necessary. One of the most enticing scenes of Dante's *Paradiso* is the "malizia santa," the sacred soldiery of Christ in the ranks of a white rose. They and angels singing as they fly–indeed the entire universe–are moved by love. The singing there and at the end of "Lycidas" display the harmony of multitudes no longer in need of rules and regulations.

Loyalty and enforced conformity might seem to have little to do with the stadium and choir effects claimed in hymning saints in visionaries, but they do. The legal system defines renegades and reduces their number. What is lawful accomplishes some of the binding that at other levels builds on instinctive confederacies. What is renegade or unlawful does much the same in strengthening honor among thieves. Confederates in crime band together much as those do who assemble legally. Their scheming has not only cohesive power but entertainment value. The manipulation of point of view in fictional versions enables the law abiding to become vicarious renegades as movies perhaps show better than any other form of narrative. *Oceans Eleven* (*Twelve* and *Thirteen*), pirate movies, *Bonnie and Clyde* (with trailers and TV series), *Butch Cassidy and the Sundance Kid, The Sting,* western outlaw pictures, and many another draw large crowds. The bonding factor on the criminal side parallels logarithms in mathematics, the inverse of exponential functions. Outlaw bands, too, have strict rules that carry the weight of ethics. Without such protocols, or in society simply manners, no orderly assembly. What is illegal one place is permitted in another, as on the high seas piracy becomes legal where it comes attached to nationalism. Blindness to wider networks is a virtual necessity for devotion to a group, one reason nationalism is strong and the United Nations not so united. The loosest and least binding of groups are those of pilgrims, refugee bands, and migrant trains. They travel with minimal regulation. Charters for overland travel in the US migration westward were good only for the duration of the journey. Tangential relations alone join unacquainted passengers on a bus or train.

For several millennia prior to inscriptions, oral transmission, to return to it, served the purposes of pre literate civilizations. Concepts of divine origins and reincarnations in India were transmitted orally. Concerning the pre writing city state of Uruk, "art and architecture combined to create an effect of power and wealth," Michael Roaf (1990) points out, "to impress the local populace and enhance the stability of the ruling group" (71). Bearing, costume, insignia, palace guards, and armies conveyed its authority without its necessarily having to post written edicts. Commands in the field were called out or sent by messengers on horse or camel back, not dispatched in writing. For the most part connections concentrated in population centers that featured a temple as well as a palace and had ritual observances as well as laws, a presumed collective history, and visual displays. Whatever the feeling of individual citizens entering an architectural monument in such a center or taking part in a ritual, it would become more or less mutual inside. Except for tourists, entering a temple, like entering a stadium, is a decisive act. Even such a casual matter as entering a home team locker room is preliminary to putting on a uniform, huddling, and running programmed plays. As on an athletic squad so in a military platoon. The uniform and insignia work their assigned cohesive force, sometimes to the point of sacrificing life.

In dissemination by scroll, collecting writings precedes declaring them sacred. The Hammurabi code, a Babylonian law code dated around 1754 (BCE), had sacred status about a millennium before the Hebrew anthology's laws and ordinances. Where a cult had gone national, not to be a member was to be ostracized. Where would-be outsiders can't be convinced, they have to be coerced. That, too, has a regrettable history as witness the Roman Empire under the spiritual guidance of Roman Catholicism. As Zagorin's (1990) history of dissimulation and hiding out in early modern Europe demonstrates, regimentation in combined civil and religious regimes adopted draconian measures of enforcement. That left non conformists just three options: to speak out and risk ostracism or worse, conform in appearance, or organize resistance. The same options apply later to ideological conformity and dictatorial modern states. In the Stalinist era, Russians by the millions found that out the hard way. Contrary articles of belief might circulate quietly in print or become outspoken. Luther's nailing the 95 theses on the Wittenberg Castle church door was a coming forth with sufficient strength to make a public stand. Where dissent incurs heavy penalties, conforming merely in appearance is a logical individual choice.

Evidence of people hiding out in any regime is necessarily scant at the time, but later revelations testify to it in some cases. The apostles and Augustine after them found the moral quandary of early Christians severe: become open and be martyred or secretive and survive. Once Christian doctrine had acquired such power under Constantine, persecuted Christians relatively soon became persecuting Christians. Even after the long trek to get to Dante's white paradisal rose we haven't forgotten the chambers of the *Inferno* where each suffers in isolation or in the company of tormentors. In medieval and renaissance Europe, nonconformists could act either on their own or as members of secret orders. The Roman and Spanish inquisitions together lasted three and a half centuries. The parties in jeopardy eventually included those who discovered such things as the nature of the solar system. Puritans who fled oppression reinstated it in New England.

Segregation and unofficial discrimination are often based on race rather than on doctrine or ideology, class, or some other form of dogma, but Indonesia, Chile, Argentina, and several other countries in modern times have put their people through right-wing ideological purges. The McCarthy era was a comparatively mild parallel in the US. In the Bush administration (continued in the Obama administration) the indefinite imprisonment of suspects without trial redirected group fear of the Red menace to the jihad equivalent. The threats in both cases were (and are) real but the reactions haven't always been prudent, as for instance in meeting the Red menace by ignoring Eisenhower's advice not to get into land wars in Asia. Arbitrary distinctions are common even in comparatively open societies. Federalists under the Adams administration in the post Revolutionary War era used the word *sedition* much as the 20th and 21st centuries have used *communist menace* and *terrorism*. Those are forceful glue words that discourage individual and subgroup deviations from a national confederacy. A tactic exercised on the basis of national security undermines it, as warfare against the red menace in Asia put hundreds of thousands in mortal danger with no demonstrable increase in homeland security with the Pacific Ocean as a barrier and a powerful navy and air force patrolling it.

Within any system of governance individual responses have a lukewarm quotient to go with commitment and rebellion. What disarms fervent commitment more than anything else is reluctance to get involved. That inertia isn't necessarily a bad thing. Falstaff's desire not to get himself killed may look like cowardice to Shakespeare, but caution in the face of a rash undertaking could just as well be called prudence, something Hotspur and Glendower are lacking. That is another reason Falstaff must be rejected by a monarch whose first major act of leadership is to invade a foreign nation. Even the highly activating powers of groupthink fail to rouse everyone. The carnage was real even where the difference to the rank and file was minimal. In territorial battles the claim to rightful possession of the land is usually based on making a selected era decisive, as in claiming Brittany (sometimes called little Brittan) Great Britain was neglecting the Roman and Galliceras.

Where strong belief teams up with civil authority, indifference offends it almost as much as opposition. The most prominent example among the ancients is again that of Israel. Up to the state's dissolution in the Babylonian exile (586), Israelite scribes and prophets reminded people of the nation's unconditional commitment to eliminating any religious practice outside that of the Yahweh cult. As Geo Widengren remarks, only one "comparatively small portion of... Israel is allowed to speak freely," whereas "the opinion of opposing portions, surely comprising the overwhelming majority of the nation, finds expression only in scanty fragments and in partly disguised allusions or in the distorted polemics of its adversaries" (S. H. Hooke, 200). The term *Israelite* itself, as Gottwald points out, refers to a "religiocultural" phenomenon, an ideal, not an accomplished fact. The northern and southern kingdoms made conflicting claims to it. In Norman K. Gottwald's terms, the two *Samuel* books and two *Kings* through *Chronicles* "grasped the fundamental point of a shared identity that, although unable to sustain political union, nevertheless endured amid the political divisions and, at the same time, endlessly complicated those divisions" (75). The religiocultural complex generated eight regicides. "Three attempts at sustaining dynasties were cut short by assassination" (76). The Davidic dynasty, though in retrospect a model of the unified kingdom, "was punctuated by five assassinations." Athalia murdered an entire royal family (76-77).

The North/South kingdoms were only the most prominent divisions. The assassinations that Gottwald mentions and the turmoil of *Kings 1 & 2* provide substantial evidence against the assumption of a unified nation. The initial indication of reluctance came at the moment of the exodus legend when advance scouts reported that the promised land had turned out to be occupied. The refusal to proceed was punished by another 40 years in the desert. That reluctance wasn't caused by humane concerns over having to kill men, women, and children and seize their land but by fear that they weren't strong enough to prevail, hence distrust of God's word at a time when so far it hadn't failed them. Difficulty in uniting regions and tribes was typical of consolidating city states into nations. That was an ongoing project in adding states and provinces in North America for some three centuries. Other kinds of union such as mutual defense leagues can be left to treaties. The Euro and the European Union are limited to economics and trade. NATO is a mutual defense league. In their freedom from dogma, contractual federations of that kind remain indifferent to other bonds that link member nations. Exits from the union don't bring civil wars. Given the complexity of the human brain and multilevel groups within groups, conflicts of interest within the mind and within each level of congregating prevent any consortium from being perfectly coordinated and at ease. Humans are incompatible with hymning white rose ranks.

Dissemination

Advances in communication, alphabet literacy, and clay tablet inscription were milestones in mind-joining dissemination not equaled again until books and still again in cinema and electronics. The age of mechanically replicated print didn't get underway until the 15th century. The invention of the radio awaited the 19th, and talk radio the 20th. Search-engine access to electronic archives came in the late 20th. Tweeting, blogging, and Facepage texting arrived about an hour ago. Messages and pictures can now be dispatched around the globe in less time than a Jane Austen character could ride over the hills to visit a neighbor. Network news falls into a weekly cycle that hinges on what the targeted audience is doing hourly. Evenings of the workweek are prime for drawing attention. The way to bury something that might cause a withdrawal of support is to announce it late Friday or before a holiday break. The effects of technology in communication on social and political orders are still being worked out in dissemination, propaganda, and group dynamics. Frauds and swindles have never lack victims, and if they collect like-minded company they cease to seem foolish to the swindled. A simplistic phrase can replace precision in thought. In a democracy the numbers game is keyed to polls and potential votes. In reducing the ranks of the informed and increasing the ranks of loyalists, one is said to have scored a public relations triumph.

Take a typical case. That fear of Saddam Hussein's weaponry was without merit in 2003 became widely recognized by the midterm elections of 2006, but by then it was too late to prevent a costly war. The Iraq military and missile system weren't capable of inflicting disabling damage at a distance, and any attempt to do so would have been addressed quickly and effectively. The much repeated slogan "fight them over there so we don't have to fight them here" had by then helped sell preemptive warfare on the assumption that if jihadists were engaged elsewhere not enough suicidal ones would be able to go abroad carrying explosives or contaminants. In effect citizens were convinced to send troops into grave danger in order to avoid an extremely slight risk themselves, a lesser risk than the hazards of travel, smoking, drinking to excess, natural disasters, fat-laden cheeseburgers, sugary drinks, domestic crime, inadequate healthcare, and a night spent in a local hospital.

They were also encouraged to ignore the intertwined complex of nations in a precarious balance when invading one of them would upset the balance in others. An intrusion of the military into sovereign territory raises resentment and enduring enmity. Inflicting what is euphemistically called collateral damage, besides inflicting traumatic mental troubles on some combatants as well as on the victims, increases the recruitment of still more extremists. Because that last drawback concerns a state of mind as well as the state of the union it isn't provable, but normal human psychology and continued retaliation in subsequent terrorist attacks argue for it. One reason for the June 12, 2016 massacre in Orlando by Omar Mateen was just such a retaliation, so Mateen said. (Some suicidal madness may merely seek a pretext.) By 2016 the awareness of what had happened a decade ago had faded enough that the same unscrupulous candidate for president thought it safe to transfer the blame to the following administration that only inherited the situation, even going so far as to claim the current president the founder of one of the terrorist groups. Vocal critics of the 2003 invasion of a foreign country weren't lacking all along, and they eventually had an effect on popular opinion, which became aware too late of the Bush, Cheney, Rumsfeld, Rice manipulation. Presented without haste and with due consideration of available evidence, an attack on a nation with a weak army, a negligible air force, almost no navy, and limited missile capacity would probably not have appealed to many beyond those who stood to profit. Persian, Greek, and Roman counsels may have been more eloquent and lucid in imperialist ambition, but the Bush administration with electronic and print media at its disposal was more effective across a wider population in a briefer time. Again the reception had to be in tune to begin with. Those on the same wave length aren't heeding other wave lengths, some of which may be rational and well informed. A similar escalation of hostilities occurred in Vietnam after JFK had decided (in James K. Galbraith's account, 2003) on a phased withdrawal to be completed in 1965. The assassination of the president ended that chance with resulting casualties estimated in allied military deaths at 282,000, in the People's Army of Vietnam and VC military deaths at 444,000, in civilian deaths (North and South Vietnam) at 627,000, in total 1,353,000. The 'one shot, a million down' syndrome would characterize other disasters, but the preliminary firing of words is really the trigger.

Exaggerating a danger is sometimes enough by itself without other distortions provided that the tempo is fast and the slogans brief. Samuel Johnson in *The Idler* was the first (apparently) to make truth the first casualty of war or as he put it in the elaborate style of the times, "Among the calamities of war may be jointly numbered the diminution of the love of truth, by the falsehoods which interest dictates and credulity encourages" (*Idler*, November 11, 1758). Thanks to the brevity of the messages and their wide transmission, indoctrination can now capitalize more easily on the dumbing down of those relying mainly on social media. To resist communal bonds requires reason's dominance of emotion, and that in turn requires training in critical thinking. Fact checking in a few doesn't offset the credulousness of the many. Whatever the period or the parties, the psychology is much the same. Some of the Romans listening to Antony (in Shakespeare's *Julius Caesar*) we can imagine realizing they are being led on, but Antony has enough status and charisma to make them stay and listen. As a subset of the extraterrestrial support that nations often claim for validation, the case is sometimes made that empire itself is destined. It looked that way in nation-by-nation shifts of power from the Near East westward. That misconception had a way of catching on along the developing edge where powers like Rome could imagine unlimited expansion. When in the colonial era it became western Europe's turn to expand, that sometimes took on Haklyut style missions. When the sea powers of western Europe–England, Spain, Portugal, Holland, Germany, and France–received that burden of empire they sent what they considered enlightenment as well as governance and commerce across the Atlantic and south into Africa. They wouldn't have linked themselves to Mongols, Vikings, and Huns, but these too marched, rode, or sailed westward. A westward march of empire would make a reasonably creditable case were it not for back-and-fill movements. In the US after settlers had reached California and the Oregon Territory, the later 19th century saw a bounce-back to the mineral rich interior. Such retrograde movements look freer of conquest than the westward migrations, but they weren't necessarily. What attracted them were resources unexploited in the first pass through. Cyrus, Darius, and Alexander had also moved east in the 1st millennium and European colonial powers south into Africa, east as well as west into India, and south as well as west into Australia and New Zealand.

Evangelical missions also went in several directions in the company usually of migrant movements and military advances. Islam and Christianity went south into Africa and in various other directions as well as westward, often at sword point. So did French, German, Dutch, and English colonies and the Arab, Portugese, and North American slave trade, the latter justified in some quarters as a master race operation. Whatever is used to gain support, nationalist expansion is opportunistic in seizing the advantage. Technology, administrative efficiency, group cohesion, and centralized power used that advantage to overwhelm scattered smaller groups and entire nations. The prominence of westward movements was basically the product of agriculture and urban civilizations developing earlier in the Indus Valley, fertile crescent, and Nile Valley than in Europe, and in Europe sooner than in much of the Americas and Africa. Migration and invasion could be characterized as the work of civilization in all its aspects, arts, industry, education, religion. Once seeds were planted in rows and animals domesticated, hunting and gathering were doomed for the simple reason that settlements and local food production concentrate the numbers, increase education, and have technological advantages, which include weapons and transportation.

Frederick Jackson Turner's essays, collected in *The Frontier in American History* (1920, 1996), are typical in playing up the advantages of agriculture in the westward migration. Each advance of the frontier furnished "a new field of opportunity, a gate of escape from the bondage of the past; and freshness, and confidence, and scorn of older society, impatience of its restraints and its ideas What the Mediterranean Sea was to the Greeks, breaking the bond of custom, offering new experiences, calling out new institutions and activities, that, and more, the ever retreating frontier has been to the United States" (38). *Democracy, freedom, opportunity, enterprise,* and *independence* select from what migrations and changes in production did: "By this peaceful process of colonization a whole continent has been filled with free and orderly commonwealths so quietly, so naturally, that we can only appreciate the profound significance of the process by contrasting it with the spread of European nations through conquest and oppression" (169-170). Technically, "peaceful process" is correct, since Turner is referring to shared authority under ordinances of 1784 and 1787. He is, however, ignoring the Indian wars and the rivalry of individuals, fur companies, migration groups, and of French, Spanish, English, and Americans variously allied and opposed. Federal support was substantial, but the advances didn't have a unified front, nor were those who got displaced uniformly nomads and hunters. Native American settlements in the Grand Canyon and four corners areas once upon a time had practiced farming and built durable dwellings. South and Central America had native urban developments, trade, farming, and urban populations to be subdued. What is missing and what one doesn't expect to find in a 19th century historian is proportionate concern with internal strife among immigrants, the forceful removal of native Americans in some places, the treatment of imported Asian labor, and other rough and tumble aspects of frontiers eventually underscored in 20th century historians. Mass migrations often illustrate these while giving them a more rough hewn countenance. Histories of European inroads in the Americas waited from the 17th to the 20th century to acknowledge the harsher side of the movement. Statements of the kind that Jared Diamond (1992) retrieves in *The Third Chimpanzee* still don't normally appear in textbooks. His roster of those who made statements scorning Native Americans includes George Washington, Thomas Jefferson, John Quincy Adams, James Monroe, Andrew Jackson, John Marshall, William Henry Harrison, Theodore Roosevelt, and Philip Sheridan (308-309), at times drawing on destiny and lofty myth uplift. Benjamin Franklin is typical in rationalizing the removal of unwanted people. In a single sentence he typifies several centuries of imperialist rhetoric: "If it be the Design of Providence to Extirpate these Savages in order to make room for Cultivators of the Earth, it seems not improbable that Rum may be the appointed means." *Providence* is the high myth part of that, *cultivators* the mode of production part, *rum* the racist part, and *appointed* the catch-all destiny part. All elements including the debauchery of the unchosen could have been paraphrased out of the *Deuteronomy* story. The cultivated understatement ("not improbable") and the implication that Indians can't hold their liquor are typical stereotypes. As a drug, rum constituted a form of biological warfare, small pox a more virulent one. Sorting out pragmatic and realistic factors from myth would require a document by document sifting, but the hefty presence of commonwealth myth in public statements is unquestionable. Because of the continued pressure of nationalism, patriotism, and imperialism, it has been slower to adjust than other areas of modernist and postmodernist concern.

The shadow contributions of illusions aren't well recognized even yet, nor are parallels between modern and ancient myths as coordination facilitators. The whole pertinent truth and nothing but sometimes comes out in cross examination but seldom where people share a national identity. The rhetoric of commonwealths concentrates on the buildup phase, which may be why lines of refugees drew so little attention until photography, newsreels, and investigative reporting took more notice. Once they did, damage control replaced silence as the tactic of choice among spin doctors.

Chapter Eleven
Organized Turbulence

Fearful Symmetry

Infectious enthusiasm works at all levels from two people to millions, and it is normally a positive force. It can also be an agent of compounded harm, however. Under illusions and indoctrination of the wrong kind, destructive causes are made to seem good or at least necessary. It is at the national level, as we've seen, that groupthink multiplies its influence exponentially, and it is at that level that indoctrination does its best and worst work. In this chapter I concentrate on group forming ideas, mostly illusions, as a source of recurrent disturbances. Historically, turmoil gone national has raised havoc in every era on record and devastated entire populations. Where brainwash is the problem, education is the answer.

First, a general observation on a chronic imbalance in the history of ideas that contributes to mistakes. The disintegration phase of buildup/ breakdown cycles tends to be under represented in common perception as well as in science and natural philosophy. Knowing what nature has in store prevents the contagious spread of a good many of the world views that have prevailed in the history of ideas. So far nothing has built up that isn't expected to break down. The evidence is everywhere. Leaves whipped about by a high wind have broken free of their anchorage and haven't yet had gravity pull them to the ground where their disarray would be reduced. They remain organized individually for a time but are disorganized as a flying conglomerate and over the winter will begin returning to the soil. That is true of unattached particles at other levels from dust in space and neutrinos on the loose to the dry sand of the desert. Some disturbances aren't subject to measurement because like the blowing leaves they don't stand still or fall into configurations. We can say that turbulence itself is organized to the extent that even at the atomic level it involves composites. Jagged forms and large debris fields are obedient to laws of physics, and it is organized societies that fight wars. At their worst in the disintegration phase disturbed human conglomerates become battlefield litter or city rubble. Armies are organized to speed that destruction. Even in their dress uniform on parade and turning heads in unison in salutation the troops are practicing the coordination that will make them more effective in the field. Without sufficient foresight the participants are more easily led into the havoc for which they have trained.

By the time the Holy Roman Empire Teutonic Knights had invaded Russia and in Sergei Eisenstein's *Alexander Nevsky* (1938) left the Russian forces no choice. The glories of conquest were built up in the knights as they typically are beforehand, and so the invasion, in 1938 heralding that of Hitler, wasn't stopped at the source. It was brainwash that set the machinery in motion and given high myth standing in the symbolism of the invaders. The thrill of battle passes from person to person and momentarily drives out the shock of the killed and wounded. The pain of the turbulence is individual, isolating as well as shocking. The glory of the encounter is an ephemeral ghost idea that passes from person to person in the buildup.

The most common words for order in its stable phases are *structure, information,* and *life,* as an integral self-propelled unit.

Comprehensive valid information, as close to objective truth as circumstances allow, normally stands on its own. Cybernetics has been the going word for such anti-entropy orders, with entropy the eventual winner. One opposite of comprehensive terms for information besides *entropy* is *chaos*, the random movement of pieces not attracted to one another and likely colliding. Others are castles in air and untested hypotheses. As one branch of postmodernism, the *science* of *chaos* (not an oxymoron) was well enough developed by the 1980s to generate productive and illuminating accounts of it, including James Gleick's *Chaos: Making a New Science* (1987) and Stephen L. Harris' *Agents of Chaos* (1990), the latter devoted to natural disasters. Some things that are patterned and perfectly obedient to laws such as cyclones seem chaotic to the areas they invade. The unpredictability of particles caught up in confusion like motes of dust doesn't negate the predictability of the mass, as individual eccentricities in populations balance each other and leave the whole broadly predictable. A disturbed mind and an upset stomach walk in concert with their owner. A cyclone takes the same predetermined spin and orbit as its hard rock planet. Geographic coordinates keep storms located wherever they go.

Nature comes up with every conceivable ratio of order and disorder. Among its formations between disorder and rigidly mathematical regularity are shorelines, fern leaf edges, and root systems. These provide visual images of irregularity without rapid change or severed connections among adjoining parts. Social networks fall in that half orderly category overall, though any given segment may be more exactly patterned and not all segments are connected by filaments. A military marching formation is exactly paced and spaced. Its deployment against another such brings a tangle of groupings and crossfire relations. That everything is constantly in motion rearranges relationships without necessarily severing them. Even humanly engineered rigid structures that last for centuries are eroding at the surfaces as they edge toward disintegration. The sands of shorelines are shifting. Mountains are crumbling and are irregular to begin with. Thanks to competing gravitational attractions within range, the orbits of satellites around stars have wobbles. Because of movement somewhere between disorder and rigid pattern, it is possible for sciences of irregularity to make methodical studies of partly unmethodical things. Any prospect offered for a social order that doesn't fit within those confines is probably not going to be realized. A natural history requiring disintegration to fuel buildup argues against any of the several exceptions posed as prospects.

This brief survey of irregularity I want to keep in mind as a context for commonwealth disruption. How lies and illusions disrupt social orders isn't subject to measurement, but it is self evident. Natural philosophy differs from natural history and its phenomena in making orderly principles out of disorder. The math that tabulates a toll in neat columns is a typical juxtaposition of messy natural or human phenomena neatly arranged as statistics, explosive reality sending bits and pieces flying but in the aftermath sentenced, linear, and paragraphed in exposition. The greatest discrepancy between *event* and *rendering* is the event labeled the big bang at the start of the natural continuum. No subsequent turbulence matches that of the first instant and the accompanying inflation instant and few acts of reconstruction equal the math and spill out of metaphors contrived to convey its nature. Not that the outset at time 0 was all chaos. It initiated buildup as well as breakdown. It isn't until a neutron and proton joins in the hydrogen atom that the history of information has something to work with, when nucleosynthesis has something to break apart and convert into something else. Four hydrogen nuclei gone, particles released, and one helium nucleus configured. Temporarily. Until its turn comes to shatter and be used to build the next element in line.

Frayed Edges

As Darwin points out (*Origin*, 2003), a plant successful in one place can become unsettled in another despite its finding a similar climate and fertile soil there: "When a plant or animal is placed in a new country amongst new competitors, though the climate may be exactly the same as in its former home, yet the conditions of its life will generally be changed in an essential manner" (175). That is true also of transplants and of species that have stayed in place and been invaded. What might function well on its own malfunctions in a crowd, bunched seedlings being Darwin's chief example among many other conceivable ones. In some things that is a struggle of roots and leafage or of tooth and claw. In people the territorial instinct and coordinating myths merge into competitive group identities. Invasions, migrations, and protection of homeland have that naturalist basis. An added ingredient with polities is their use of indoctrination for collective defense and aggression.

Animals too of course have pecking orders as tall trees take the light seedlings need, but social orders are exponentially more complex in the awarding of offices and honors under the ruling hierarchy. Domestic affairs are subject to ranked offices similar to military ones. The sources of discontent that disturb them from within are too many to be indexed briefly, but violated civil rights and injustice are among the common ones, enforced conformity in beliefs another. Theocracies trigger additional friction whenever doctrinal enforcement becomes too draconian. Parties seeking to overthrow such a civil/religious combination use similar appeals to high myth on the grounds that if, for instance, the Holy Roman Empire claims divine sanctions, Reformation movements can scarcely do less. If a Stuart king and Anglicanism asserts them, Puritan rebels will too. The war of words gets underway long before the use of weapons does, and it is seldom scrupulously fact observant.

In an early example Mami's instructions to Ninurta for the destruction of Anzu raise pre battle rhetoric to the intense level that will typify adversarial harangue in other instances, but in this case without a civil or even an expressly human component:

> Muster your devastating battle force,
> Make your evil winds flash as they march over him.
> Capture soaring Anzu
> And inundate the earth, which I created—wreck his dwelling.
> Let terror thunder above him,
> Let fear [of] your battle force shake in [?] him,
> Make the devastating whirlwind rise up against him,
> Set your arrow in the bow, coat it with poison. (Dalley, 212)

Outbreaks of that kind in ancient narratives anticipate by several centuries the exhortations of Hebrew prophets claiming to represent God's will. The animosity could just as well have been projected into Minerva, Zeus cursing Prometheus, or Gaia rallying the offspring of Kronos (in Hesiod). For that matter, it could be Pope Urban II inciting the crusades, Cromwell the Puritans, or Robespierre the French. That nothing supernatural need be inserted is clear in arguments like MacArthur's for defending South Korea: "I can almost hear the ticking of the second hand of destiny. We must act now or we will die" (cited from Martin J. Medhurst in Hogan, 161). A recent candidate for president of the US was filmed saying that if his opponent was elected it would be the end of civilization. That the most blatantly false claims generate a following is further testimony to the stadium effect. Of all the components of harangue, truth is the most expendable.

Both realm-supporting illusions and rebellions easily become idea epidemics of the kind Daniel Kahneman features in *Thinking, Fast and Slow* (2011), in that case the result more of a mental process than of a clinical malady, although the two overlap. Thinking deliberately and by the numbers is more burdensome than jumping to conclusions and much less contagious. For other ancient examples recall those mentioned earlier, the Akkadian Sargon (2270-2215);

his son Rimush (2334-2270); the Assyrian Tiglath-pileser I (1114-1076), who initiated some 500 years of Assyrian turbulence and dominance; the two 9th century Ashurnasirpals; Shalmaneser III (858-824); Tiglath-pileser III (745-727), who murdered his way to the throne; Sargon II (721-705) and his son Sennacherib (705-681), a ravaging tyrant of great cruelty; Ashurbanipal (668-627); the Assyrian Adadnirari I; Shalmaneser I; Tukulti-Ninurta I; and the Amorite Hammurabi. These were among the noteworthy imperialists about whose rise and fall historians and archaeologists have enough information to make judgments. For the inner workings and schemes of dynasties we have Shakespeare's histories and tragedies to consult in a dominantly secular frame of reference and the Hebrew anthology for authority vested jointly in kings and prophets. That not much has changed in that regard is clear in madman documents like that of the before-mentioned Sheikh Abdullah Azzam in "Martyrs: The Building Blocks of Nations." That history is written in blood and achieved with lofty edifices of skulls has a glaring fallacy: history isn't exclusively written by martyrs and victims and seldom if ever constructively.

    It would of course be a stretch to say that tumult in social orders in any way derives from competition in the plant and animal kingdoms in general or from turbulence in far places, but they are parallel, and an encompassing realm does influence its parts. Human psychology wouldn't have developed in the way it did but for what the environment selected by the usual means of population thinning. A rough world and a hard life over time have an effect. Where non sentient disruption creates wreckage and debris and where species produce populations in excess, it is probably unrealistic to expect one species however sapient to be immune. That was a less tangled matter in clans and chiefdom tribes than it became in urban civilizations, but the emotional brain is the same in whatever setting. Its easily inflamed mental fever may simply spill over from quick trigger reactions to perilous situations. These are less well suited to a crowded planet. Darwin, citing Malthus and De Candolle's war of nature, makes a similar point in several places, one of them "Natural Means of Selection" in the 1844 Essay preliminary to *On the Origin of Species*. Grazers and browsers with abundant forage look to be exceptions to survival feistiness as do non-predatory sea creatures, but not much is immune to peril even if it isn't involved in predation flight or fight. Competition in mate selection affects grazers and browsers as well as prey and predators. Defensive strategies include deceit, camouflage, and striking first. That most people are compatible and reasonable most of the time is no minor achievement under those circumstances.

Strife among opposed parties with local commitments takes a chronic geographic form in city-country and center-periphery tension. That added to trouble in systems of rule, belief, and modes of production. Urban centers have the advantage not only of concentrated numbers but of wealth, an advantage that dissipates with distance and in combination with the territorial instinct triggers turf battles. In antiquity when cities that had formerly been regional centers fell into satellite status they were more likely to organize rebellions against an imperialist dynasty than scattered and thinly settled regions were. Partly for that reason, inhabitants of overwhelmed cities were often either eliminated or transported. The Babylonian exile of Hebrew leaders was one of many such episodes. Frontier disturbances are the products partly of distance but also of the fact that it is more obvious away from a ruling center what its pretensions are. Rebellion is more likely on the fringe attacking inwardly, aggression more concentrated in the center attacking outwardly. Like the 'high places' Josiah sought to close down (*2 Kings* 23:19), the remoter areas are more subject to outside influences.

Power concentrates heaviest where palace and temple join, but even less temple-dominated urban centers Fernand Braudel (1981) finds functioning "like electric transformers." Despite being built up of diverse elements, they are concentrated enough to "accelerate the rhythm of exchange and constantly recharge human life." They are "born of the oldest and most revolutionary division of labour: between work in the fields... and the activities described as urban"

(479). Lines of communication and trade routes pass through or end in such centers. Administrative bureaucracies concentrate there unless foresight places a capital in a lesser urban area outside a major one, as is the case in several state capitals in the US. That same concentration accelerates the spread of the mental fever that turns an urban center imperialistic where it is also the seat of government.

As to upheaval on the frayed edges of nations, history gives us many accounts of how it plays out. The reach of imperialistic power since the beginning of colonialism places the troubled frontier half way around the world. Since 1898, the US has engaged in approximately nine significant conflicts plus Marine Corp landings, the 1898 Spanish American war, the two world wars, Korea in 1950, Vietnam in 1964, Cambodia in 1970, the Persian Gulf War in 1991, Kosovo (with NATO), 1998), Afghanistan in 2001, Iraq in 2003, with miscellaneous other drone and air strikes. Syria and Lybia are two of the current problem areas. Some of the century's conflicts have had only peripheral connections to theocratic myths or imperialism, but War II was anti-Semitic in Germany and other European countries, and religion has been a sponsor of skirmishes and terrorism through much of the history of Palestine. Foreign policy in several western countries had ethnic and religious implications for Jewish immigration, especially in War II and after. An Arab/British agreement that limited the number of immigrants made Israel hesitant to fight on the British side even against Germany. The Jewish occupation of what had recently been Arab territory underlies the Abdulla Azzam-Osama bin Laden terrorist movement. Jewish/Christian/Muslim distrust, incidental in the Persian Gulf War, was more prominent if never fully explicit in 2003. In the prolonged war in Afghanistan it has been equally dogmatic and theological at bottom in its Taliban elements. In terms of Russian, US, and China spheres of influence, the disputes are usually at the perimeters, as were the communist/capitalist underpinnings of the Korean and Vietnamese wars. Building walls is generally limited to smaller areas but has been tried on national borders as well. It is meant to limit traffic and prevent infiltration and thereby limit conflict on the Robert Frost principle that good fences make good neighbors ("Mending Wall"). How well that works on a national scale depends entirely on the parties involved, how they are armed, transported, and motivated. The most spectacular achievement in wall building by common accord is the Great wall of China. Least effective are those that divide families and markets like the Berlin wall and sections of the US/Mexico fence. Every border crossing with a guard station is a variant. National sovereignty is the issue. Barriers help prevent conflict where indoctrinated citizens increase divisiveness.

Even in modern democracies hospitable to nearly everything, sectarian tension remains alive in the mutual distrust of Hindus, Muslims, Jews, and Christians. The friction between Pakistan and India has a residue of ancient Muslim/Hindu conflict, and again territoriality plays a role. Doctrine and ideology overlapped with frontier confrontations in much of the 20th century turmoil. When ethnic and national concerns are added, the issues become extremely murky. The Third Reich was a prime example in which megalomania, anti Bolshevik, anti Semitic, nationalist, and racial matters were in the mix along with German imperialism. The Goebbels, Dietrich, Hitler propaganda machine used a high volume, puffed up discourse to glorify the cause.

Motives for aggression have an obvious bearing on national identities and the wars considered in Phillip Knightley's *The First Casualty* (1975) and Margot Norris' *Writing War in the Twentieth Century* (2000). Piero Scaruffi (2009) lists 121 wars, insurrections, and genocidal incidents in the 20th century plus a small slice of the 19th, and 366 major conflicts from 1740 to 1974. In a sizeable share of these either ideology or doctrine played a significant role along with the center/perimeter tension of imperialism and of course the territorial imperative and usual hegemonic issues. Evidence suppression and denial are among the common means of pre battle persuasion along with territorial nationalism. The doctrinal contributions were greater earlier and ideological ones later. The

Crusades, the Reformation wars, the Muslim conquest of the Iberian Peninsula in the 8th century and the *reconquista* completed in 1492 were holy wars with a territorial base and advancing fronts under dynastic monarchies. The papacy viewed the Iberian reconquest as the first crusade. Dictatorial regimes are similar, sometimes working without needing to conceal brute power. By mid 20th century the embattled frontiers were mostly ideological. By the end of the century they had shifted back to retrograde doctrine-oriented terrorist movements less defined by national boundaries and urban ritual centers.

Coerced Commonwealths

The weakness of imperialism lies in the strength of the imperial power. That isn't the paradox it may seem. Empires gain and maintain control by force and by that same means raise the antagonism that weakens them. Phoenician and Athenian naval power overcame Mediterranean dispersal in geography and culture for a time, but in the case of Athens even at the height of that power allies were prepared to bolt. That isn't to say that all colonies react in the same way. The 13 American colonies and India rebelled against England. The Australian, New Zealand, and Canadian provinces did not. Carthage grew too large and independent to stay attached to Phoenicia. The USSR divided into multiple nations within a few decades of their union, and an oppressive Moscow central was one reason. As R. N. Berki observes concerning Greece (1977), following the Trojan War (around 1200 BCE) lesser states broke away from consolidated ones, as many as 200 with populations ranging from 10,000 to 100,000 (42-43). In the colonial era, South Africa was another trouble spot distant from the Dutch and English centers of colonial rule. The return of captured territories to their former independence testifies to the combined persuasion of ethnicity and geographic separation and to inherent local hostility to imperial powers.

No group tries harder to recapture lost ground than an ideological or religious one that casts territorial possession into holy land terms and enlists martyrs to recapture it when it is lost. That along with the exodus myth indicates how overlapping the territorial and groupthink instincts are. When homeland becomes God's country almost any radical extreme is possible. Under the best of conditions it isn't easy to find a balance between center and periphery, as witness the recurrent tension between federal authority and states' rights in the US, generally speaking one of the more successful arrangements under the Constitution and Bill of Rights. In eastern Europe, Latvia, Estonia, Belarus, Bosnia, Lithuania, Turkmenistan, Uzbekistan, Kazakhstan, Serbia, Bulgaria, and the Czech Republic regained a degree of their former autonomy with the breakup of the Soviet Union, formerly bound together under ideology and iron-hand rule that by means of imprisonment, exile, and execution eliminated dissent real or suspected.

The initial American colonies were willing to risk a great deal to break free of monarchical England in one of the remarkable rebellions against a distant administration. To succeed, the colonies first had to overcome rivalries among themselves. Although they did so to a significant degree, Alexander Hamilton doubted that their divisiveness could be completely eliminated. A breakaway commonwealth incurs new or renewed divisions of its own: "A man must be far gone in Utopian speculations," Hamilton warned in *The Federalist* (1961), "who can seriously doubt that, if these States should either be wholly disunited, or only united in partial confederacies, the subdivision into which they might be thrown would have frequent and violent contests with each other. To presume a want of motives for such contests as an argument against their existence, would be to forget that men are ambitious, vindictive, and rapacious" (108-109). That throws politics back on inbred territorial psychology and the self centered individuality of organisms. They collect into groups for various reasons, but each member remains an integral unit. On a sparsely populated continent, the new federal union soon became imperialist itself, legitimately with the Louisiana Purchase, so it seemed from one perspective.

Pause a moment on that example with its Constitution and the Civil War breakdown. A saving feature of the initial federation was lack of ideological articles of belief dominant enough to require compliance. Protected minority rights and religious freedom were essential to the agreement. Liberal political thought assumed that ordinary law and order could hold vindictive and rapacious elements in check. All rule is hierarchical, but American hierarchy had two wrinkles: the hierarchy of wealth didn't coincide with that of rule (more so lately than earlier), and people rotated in and out of office regardless of lineage, thus disallowing dynasties. Thanks in part to economic policies that dispersed wealth, those two variants of hierarchy eased the pressure that would have increased with any return to a durable lords-and-serfs version. In the homestead acts, family independence found not only legal support but economic backing. Instant and lasting middle class standing came of owning and improving on homestead acreage. The tension among ethnic elements and between states and the federal union was relatively mild early on despite the problem brewing in slavery, which had to consider people beasts of burden or deny the Bill of Rights. The southern version of lords-and-serfs in slavery was too fundamentally at odds with the democratic project to last forever. The leisure class in Europe transported its privileges to other parts of the new world as well, largely English and French in North America, Spanish and Portugese in Central and South America, and Dutch as well as British in South Africa. In North America, however, they encountered a landed and mercantile middle class with territory and resources. Cattle barons were imaginable in Montana and robber barons elsewhere but the equivalent to nobility only in the South and only for a limited time.

In the Civil War, fictions and stereotypes took their familiar course. Southerners searched out biblical passages that in their minds reinforced the class separation. Troops on both sides were all the more willing to fight in being assured they would land in heaven if slain, a supporting illusion instilled by cult rather than by nation. Point by point negotiations might not have guaranteed avoidance of war, but inflammatory myths at play made certain that it would be war without them. Patience might have produced a phase-out plan or a wage-labor agreement as an economic proposition. That wouldn't have removed the illusion among some that a divinely ordained natural order created an inferior species to harvest tobacco but it might have disarmed enough partisan spirit to prevent the war. Or the North might have tolerated a separate southern union, since nothing declared North America to be suitable for just two nations, Canada and the US. Conceiving of the continent from scratch might have had it unified under a loose federation with regional rights or divided into several nations.

Almost any negotiated agreement for shared power would have been better than what actually happened in the Civil War and in border wars among the French, Indians, and English. The Revolutionary War, too, to go back to it, was the product not only of patriotism and territorial instinct but of English imperialism. From the far side of the Atlantic that could have been a negotiated matter with the easing of central dominance and apportioning of regional sovereignty. Separately ruled nations do coexist in restricted areas, witness Wales, Scotland, Ireland, and England. Witness Sweden, Norway, Denmark, Finland, and Lithuania. Full participation in the union proved to work in the new states, two of them, Alaska and the Hawaiian Islands, at a distance and the Philippines as a satellite until after War II. Several territories fall under US administration but are relatively independent, American Samoa, Guam, Northern Mariana Islands, Puerto Rico, and the Virgin Islands. In the wisdom of hindsight it seems likely that slavery would have ended on its own in the South as it did elsewhere and with it the slave state/free state hostility. The contrast between prewar rhetoric and postwar grief fell into the age-old pattern, first high emotion, then, too late, reason and regret. The proportion of truth was higher in the aftermath than in the lead up.

The Use and Abuse of Discourse

In the Civil War brain fever dressed in blue and gray raised havoc on a horrifying but not unusual scale. Establishing rational controls ahead of time is the best deterrent, and it is assisted by shedding illusions. Democracy in and of itself isn't an answer, but the settled constitutions on which it is founded can be. As 5th century Greece illustrates, generating exceptions to oligarchy needed special conditions. The Greek exception came in an era of nearly universal monarchies and oligarchies. The constitution of Solon and the assemblies that followed were the product of far-flung populations in mountainous, sea-carved territory. As Antony Black points out (2009), the relatively self-contained regions tended to be small and their citizens participatory. The practice of holding public hearings and debating matters of mutual interest carried over to Athens. Bloodless parliamentary skirmishes took the place of frontier battles in part because mountains and bodies of water separated the settlements and cities. A mostly naval power wasn't in a position to dominate every sovereign power, though the closest one, Sparta, feared just that from Athens and received no assurances to the contrary under the imperialist policy defended by Pericles. Rhetoric as a formal discipline taught in academies flourished under those conditions. Disputation and complaint were ancient but didn't take root in institutionalized training until 5th century academies in sophistry and philosophy. Rhetoric in that framework was expected to be argument. Contested points developed means of formal presentation under procedural rules. That reduced the chance of either illusions or proclamations sweeping through an assembly, though they still did.

Where assembly representation gains a footing like that, a prominent reason is the self interest of individuals and self-identified groups that insist on being heard. It also depends on willingness to listen. The North/South American debate had just such a forum but not the listening and bargaining. Generally speaking, differences are handled by just two means, suppression and negotiation, but a third means is the endorsement of a common factor such as resistance to an external enemy or a commanding common illusion. The tribes of Israel had both, accompanied by the expectation of conformity in belief. Even so some citizens were more independent than **D**[euteronomist] and **P**[riest] suggest in texts written long after the fact. As both archaeological and textual studies reconstruct Hebrew history, it featured on-going struggles between outlying districts and national hegemony. D and P proposed a common heritage based on external enemies and the legendary Abraham/Yahweh covenant. In the hands of prophets and priests that wasn't a negotiable point. That such a presumption was illusory as many such before and after have been didn't prevent it from exercising considerable cohesive power. What might have been practiced in households everywhere isn't determinable, but the texts imply that it wasn't uniform outside of Jerusalem. Maintaining a united state surrounded by dominant powers proved to be impossible on several occasions. The factions dominate in *1* and *2 Kings*. Despite the unifying factors of common enemies and common beliefs, the tribes, districts, and their exposure to other customs remained divisive.

Even in times of national emergency when the rhetoric of sub order alliances and factions fades in favor of a common cause, every large population has subdivisions. For Hamilton the fundamental cause of social disturbance wasn't this or that weakness in government or a given administration but "the nature of man" himself. Ambition in politics and zeal in religion have ever "divided mankind into parties, inflamed them with mutual animosity, and rendered them much more disposed to vex and oppress each other than to cooperate for their common good" (131). *Ever* is perhaps too strong given the successful consolidation of territorial states that combine multiple urban areas. People will join a federated union of their own accord if looks to be the right one. As Howard Bloom (1995) observes in confirming the Hamilton principle, however, along with the Lucifer principle in literature, psychology, and philosophy, "from our best qualities come our worst. From our urge to pull together comes our tendency to tear each other apart. From our devotion to a higher good comes our propensity to the foulest atrocities" (3). That list of paradoxes could be extended. From high principles comes high dudgeon, from idealism cynicism, from high-mindedness high-handedness. Burning someone at the stake or engineering a massacre requires a powerful mind-drug. That begins with parents and children and extends to cities, counties, states, political parties, and federal unions. Facts make powerful unifying forces as well, and nothing prevents their forming combinations with fantasy. In a sound judicial system they lead to verdicts of innocent or guilty and in national situations to policy decisions, sometimes armed response to aggression. Facts join with inference in conclusions based on convincing but not certain evidence, as stereotypes mix fact, surmise, and fiction.

    Lineage-based rule seeks a cure for potential anarchy by settling the question of executive descent ahead of time, but that too can backfire. When in-house feuds result in palace revolutions, something meant to coordinate ends up dividing. Xenophon's Greek 10,000 were hired by the Persian Cyrus to take the crown from his older brother. Shakespeare's Richard III (but probably not the real Richard III) disposes of various rivals on the way to the English crown. Hamlet Senior, to mix fictional with real examples, is murdered by his brother. Two of Lear's three daughters are viciously competitive. Statius' *The Thebaid* sets the sons of Oedipus (Polynices and Eteocles) against one another over control of Thebes. In democracies, partisan disputes replace such deadly in-house feuds, but like other vested interests they too can heighten us/them differences to the point of paralysis, as the Republican Congress set out to do in the acrid atmosphere of Washington, D.C. from 2008 to 2016. The main advantage in democracies is that changes usually come under a constitutional system that heads off internal trauma. Losing an election isn't the end of the world.

National administrations, family, regional differences, and cult loyalty can pull individuals in several directions at once. The down side of idea-based adaptability and flexibility is idea based maladjustment. Entrenched ideological and doctrinal factions set the rest of an open, democratic population against them. That is more markedly the case where church and state aren't sufficiently separate and conformity is enforced. Hobbes and Spinoza sought to heal a common rift between civil and ecclesiastical authority by giving the chief magistrate final say on religious orthodoxy, a sure way to enlist another Cromwell in another rebellion. The purges of disagreeable elements in fascist, Nazi, and Communist regimes came of intolerance toward deviations from the party line.

Sir Thomas More is a celebrated case of someone caught between rigidly programmed parties, the crown with beheading authority and the Vatican presumed to hold the keys to eternity, both drawing on what they considered to be the natural order. It would have done little good to point out the entirely conventional nature of both the monarchy and the Vatican. Each was a contrivance maintained by indoctrination and put on display in pomp and circumstance, choreographed in stately slow motion and in costume. Neither was compatible with an objectively assessed natural order, nor did the actual reigns of either bear much resemblance to its pageantry and symbolism. More was executed for choosing one set of illusions over another. Despite the examples available in topography, climate, and vulnerable forms of life, no critiques based on the actual universe were in circulation. The base line of fact was missing. Miscellany, disorder, and irregularity were systematically explained away by apologists for prominent doctrines. More's contemporary polymath

Copernicus lived in Prussia, and Galileo came almost a century too late. Both insisted on an important astronomical fact, the heliocentric solar system, that by implication challenged much more.

Polities with and without a Homeland

To the already extensive literature on the rise and fall of empires and nations, an account of their myths and illusions has only footnotes to add except for the effect master illusions have. The rises and falls of bronze and iron age civilizations came largely in technological and material terms. Food and water supply, availability of metals, climate, and population numbers were the critical items. Making the mode of production primary has the advantage of being based on more or less measurable quantities. Fantasies and other psychological matters in contrast are intangible and have to be judged by their repercussions. The combination of fact and fancy is further entangled in leadership falling to good, bad, and indifferent dynasties. The irregular rise and fall of nations itself needs no argument except to underscore the function of organized deluded groups within them. A nation can perish quickly or last for centuries. The Elamites and Hittites, to take an unusual example, persisted through much of the period from Uruk in the 4th millennium to the Persian empire. A dynasty that prolonged is exceptional. Sumerian and bronze age Akkadian, Babylonian, and Assyrian regimes came and went for the reasons nations normally do–exhausted resources, population problems, foreign invasions, and over-extended ambition. Administering an extended territory under conditions of slow communication and transportation added logistic problems. Where throughways expedited trade they also allowed the movement of enemy cavalry and war chariots. The land powers–Assyria, Babylon, Persia, Egypt, and later Sparta–lacked sea power. The sea powers–Phoenicia, Carthage, and Athens–lacked land power. The competition any dynasty encounters on the edges keeps it from making indefinite extensions of the kind Herodotus' Xerxes envisions for Persia and other empires imagine in their formative years.

Moderate size expeditions tell us in compressed form something about social orders momentarily free of ancestral myths. On a scale just large enough to serve is the Persian expedition of Xenophon's Greek 10,000 in *Anabasis*. Enlistees in the expedition had a minimal heritage in common but a strong pragmatic cause in confronting perils. Their coming together as a mercenary corp suspends individual prior commitments to family, gens, region, and city state. The march home after they abandon their initial mission holds them together, and its successful completion allows them to disperse as they wish. They come together initially as mercenaries bent on dismantling a Persian regime, stay together for survival, and disband. They thus make a compact if simplified model of polity assembly and dissolution, in this case largely free of illusions. What brings the corp together minimizes their differences for the duration of the expedition without any notable reliance on propaganda enhancements. That makes them an exemplary if minimalist example of how a social order can function free of myths. Their credentials as a provisional polity go something like this:

- Anthropologists sometimes designate 10,000 as a minimal number to generate the division of labor and wealth that makes up complex groups. An army corp with officers, peltasts (shield-bearing and javelin-armed troops), archers, hoplites (more heavily armed), and cavalry has enough specialization to illustrate ranks needing to be integrated and synchronized.

- On the move the contingent encounters a succession of hostile and friendly foreign powers that over a two year span exposes it to what a nation might face in a more extended period. Diplomatic flexibility and ingenuity are tested on those occasions if not with the complications of nations competing in trade and having border disputes.

- In internal organization and at border-crossing confrontations, the leaders encounter not only external hostility but internal anxiety. As a historian and a leading figure in the expedition, Xenophon is one of the negotiators.
- That the leaders are themselves answerable to collective judgment makes the group a traveling judiciary of sorts. Though lacking a constitution or other explicit charter, they have comparable precedents in a military code of conduct. Provisions and marching orders are administered from the top down without legislative sessions.
- With a daily need to find shelter and provisions especially in the mountains in wintertime, the troop faces decisions as to raiding, plundering, backing off, negotiating, purchasing, or stealing. How they should present themselves to those they encounter is a recurrent diplomatic question. The feints, movements in the dark, and diversionary displays serve as exported propaganda in what could develop into hit-and-run warfare.
- The disbanding gives the narrative the impression of a finished story typical of other migrations. The dispersal is the phase of the buildup/breakdown cycles that often prevails. That the components go off to join other social orders is the way of dissolved or absorbed polities, indeed of everything under the universal buildup/ breakdown/buildup pattern beginning with hydrogen neutrons and protons.

The Greek 10,000 returning home is largely a success story of cooperation, dedication to a task at hand, and disbanding without rancor. It is a turbulence dodging episode. As opposed to most other migrations, the *Anabasis* contingent isn't looking for a land of its own and doesn't have to harass any of the populations it encounters. The commitment of its members from the outset is provisional and temporary. It is free to rely exclusively on pragmatism and realism under the organization it had in its original mission. Polities that can absorb foreign influences likewise often escape the worst divisions of regimes stiffened by inherited doctrine and enforced dogma. Thanks to its population from the outset consisting mostly of migrants, the US is a prime example of unforced beliefs under a separated church and state. That its states, counties, and municipalities don't coincide closely with ethnic groups makes for complex regional patterns. The identity of any given citizen combines family, social circle, ethnic, regional, and federal union elements besides whatever comes attached to an occupation. The number of allegiances is part of the flexibility that fits individuals into a national aggregate.

Or doesn't fit them in as the case may be. Consider Reginald Hill's putting heroic codes similar to those of epics and of *Anabasis* under a woman's scrutiny in *Arms and the Women* (1999). In that view, epic accounts of expeditions overlook too much

> *Arms and the Men they sang, who played at Troy*
> *Until they broke it like a spoiled child's Toy*
> *Then sailed away, the Winners heading home,*
> *The Losers to a new Play-pen called Rome,*
> *Behind, like Garbage from their vessels flung,*
> *—Submiss, submerged, but certainly not sung—*
> *A wake of Women trailed in long Parade,*
> *The reft, the raped, the slaughtered, the betrayed.*
> *Oh, Shame! That so few sagas celebrate*
> *Their Pain, their Perils, their no less moving Fate!*
>
> *But mine won't either for why should it when*
> *The proper Study of Mendacity is MEN?*

As this indicates, it pays to look behind histories for those left in the wake. They may be a source of future Suffragettes. The more tightly bound a formation, the more it is likely to leave out, as a cult excludes all but members and an ideology sets itself against rivals. One wonders what changes in the Greek 10,000 would have been necessary had subordinate or co-equal women been included instead of waiting at home.

In the interests of coherence, historiography and fiction make similar reductions in the ranks, as much of the population of Thomas Hardy's rural and village settings is unlisted in any of Jane Austen's casts of characters. The roster of voters had notable missing elements in the US until the 20th century was well along, until August 18, 1920 to be exact for women. Though the last of the Reconstruction Amendments passed in 1870, it wasn't until the Voting Rights Act of 1965 that racial discrimination in voting was prohibited. Over 10 million illegal immigrants currently fall into that category. Historiography did however eventually catch up to the deprivation that had often been a source of turbulence, including uprisings and revolutions. In adding to data collected by P. A. Sorokin in *Social and Cultural Dynamics,* Zagorin's account of armed insurrections (1982) between 1500 and 1660 finds over sixty in Spain, France, and England. The volume swelled once the printing press made the wider dissemination of polemic materials possible. England in the early to mid 17th century saw an increase in publications agitated by social issues, doctrinal differences, and privation. Who could vote was one of the recurrent issues.

The need for a traveling group to stay together and the provisional nature of the bond are usually enough by themselves to postpone the breakup phase until the appropriate moment, but rules and regulations can be set beforehand. In most cases of a traveling, provisional social order, dissent or mutiny on the part of the shanghaied or drafted is treated harshly. Conscripted members of a marching order, like impressed ship's crew members, make for a complex integral unit. In the War of 1812 Americans were seized for crew members of English ships and when war broke out found themselves on the wrong side. That was ironic in view of the fact that a prime cause of the war itself was impressment.

Rank

Stature earned by achievement is honored enough to help coordinate social efforts, but lineage based status granted by convention is another matter. It depends largely on convention and ceremonial reminders such as royal progresses. It simultaneously keeps order and raises the resentment that leads to future turbulence as witness the several revolutions on the way from monarchies to democracies and communist regimes. Cities like the 18th century Venice and Naples that Braudel (1981) portrays featured durable assignments of rank as most 18th century European dynasties did and some still do. It took fifteen years of residency in Venice to crack the outer circle and twenty five, presumably on one's best behavior and with wealth, for full citizenship (518). Democracies have durable divisions of their own, as racial segregation in the post bellum South had custom and some legality behind it. These don't always bring turbulence, but under strain they can. Placing few at the top and many below necessarily characterizes administrative structures, and a functional ranking of that kind is easily confused with status outside the organization itself, as a five star general has carry-over social rank outside the army. As different parties come into and leave office an administrative hierarchy transfers defined and regulated powers. That temporary status becomes blurred, however, where entitlement from one sphere bleeds into another. Rank and wealth both travel from setting to setting, partly because display carries over in clothing, manners, and the vehicle one drives.

Examples are familiar enough to need only a reminder, like the snobbery of Plutarch (1960) concerning how Roman patricians regarded inherited station: "a man who occupies himself with servile tasks proves by the very pains which he devotes to them that he is indifferent to higher things," an oft repeated statement. The appreciative patrician observer of a statue of Zeus or Hera thereby ranks higher than the plebeian artist who carved it despite the first merely casting an appreciative eye after the sculptor has exercised imagination and skill. A great poem that charms us doesn't mean to Plutarch that the poet himself is charming (166). As the Italian diplomat Baldassare Castiglione (1479-1529) suggests in *The Book of the Courtier*, if a courtier as a beneficiary of patronage takes up art or poetry he should do so casually to show that achievement in any endeavor (stature) is incidental to rank (status). Making status even slightly dependent on performance calls the ranking system into doubt. Transfer from one to the other as in the knighting of someone must have the blessings of authority and the sanctions of a proper ceremony.

In England, artisans and tradesmen were objects mainly of satire until Thomas Dekker, though Chaucer, William Langland in *Piers Plowman*, and More's *Utopia* were exceptions, as was any pilgrim treated as a soul rather than a mason or baker. Common people weren't featured protagonists until Defoe. Craft or occupation, character, and lower class standing didn't receive detailed treatment in fiction until Dickens, Gaskell, and Hardy. Of all literary forms, the novel, because of its interior reporting and management of conflicting points of view, is perhaps best suited to handle the tangled relations of individuals in a rank-conscious society or other set where social relations are based on status conventions. The rebellious child, a soldier who fires into the air rather than at the enemy, a team member who doesn't go all out are transparent to an omniscient narrator. They may need only a stern parent, an unpopular manager, a drill sergeant, or a domineering boss to set off their mutiny. Where the ranks are fixed by lineage, the consuming preoccupation of the upper echelon is with their comparative standing. Where those assigned lower rank don't adjust in deportment, the result isn't normally full scale turbulence but one of the frictional rough patches that raise tension. It is the constant irritation of low-level disturbances that build into the democratic and Bolshevik revolutions behind polemical speeches and publications.

Also of interest to novelists is ideological alienation. The novels of John le Carré and Alan Furst portray defectors and double agents caught in an international crossfire that often leaves them stranded. Operatives in the field can't fully trust even their own colleagues in the spy trade. Furst's *Night Soldiers* and *Dark Star* are prolonged nightmares of disintegrating NKVD and OSS predation, fascist and communist maneuvers, and British and American agents caught in a crossfire. If we had no other account of fractured allegiances in the wake of tzar and king displacements Carré's *Our Game* (1995) would serve. Chechens, Armenians, Kazakhstans, Jews, Ossetians, Cossacks, Sheikhs, Murids, Georgians, the Inguish, spymasters and accomplices, the NKVD, and American and English intelligence are, each and every, pitted against all others, dragged along by assumptions of dubious merit. In general the capitalist/communist contention is over the standing of the middle and lower classes translated into management and labor. The accumulation of property and wealth characterizes the bourgeoisie, which has to take in excess to have funds to divert into investment and property management.

Bacon couldn't foresee the threat that the sciences posed to prevailing means of gauging status and of rule, nor could he or anyone foresee the threat to monarchy that would come from an enlarged middle class engaged in trades, manufacture, and eventually large scale mechanized agriculture. The connections were too distant to be foreseen, but in retrospect the sciences led to the technology and the technology to machines, devices, and specialized mechanics and operators. The employment helped empower the masses and dethrone kings and queens, the masses at first joining middle and lower class, as the English revolution was led by middle class spokesmen but had laborers and unemployed in the ranks. In the 16th through the 19th centuries, every harnessing of new energy was another step not only toward the elimination of hand labor and a hunter/gatherer way of life butwealth's displacement of lineage elitism.

Thomas Jefferson put a phase of that in terms of progress as most people did. Returning eastward as Lewis and Clark did, a traveler would first see natives "in the earliest stage of association living under no law but that of nature, subscribing and covering themselves with the flesh and skins of wild beasts." Such a traveler "would next find those on our frontiers in the pastoral state, raising domestic animals to supply the defects of hunting. Then . . . meet the gradual shades of improving man in cities and sea-coast towns supplied by trade" (1993, 211). Within a few hundred miles, such a traveler would thus be crossing centuries of social change based on trade, agriculture, and manufacture. With minor adjustments, almost anyone in the 19th century could have written as Jefferson did, "I have observed this march of civilization advancing from the sea coast, passing over us like a cloud of light, increasing our knowledge and improving our condition" (211). *Knowledge* and living *condition* have little if anything to do with the form of government in this brief summary, but Jefferson was overseeing early stages of the distribution of wealth and of agriculture based on multiplied occupations and the homestead acts.

The oversimplifying of barbarity is clear in Jefferson's observation that the latter came "under no law but that of nature." The stages of progress in his account were roughly those of Morgan's savagery, barbarism, and civilization, Malthus' hunter, shepherd, and farmer, and not altogether different from the roughed-in archaeology of Thucydides and Lucretius. Morgan proposed American Indians as an example of the first two phases, in his case without overlooking Iroquois customs and tribal rule. Largely missing in that version of progress are 19th century mechanization, smoke stacks, high finance, coal mines, and railroads in the hands of robber barons. Once plow horses enter a region, buffalo hunts are on the way out. Once a tractor enters plow horses in turn are on the way out except in those determined to stay in the 19th century. With tractors steal and iron come as the supporting industries.

Even at Jefferson's comparatively early stage, another side of progress was already challenging the optimism. In Jefferson's words: "In truth I do not recollect in all the animal kingdom a single species but man which is eternally and systematically engaged in the destruction of its own species. What is called civilization seems to have no other effect on him than to teach him to pursue the principle of bellum omnium in omnia on a larger scale, and in place of the little contests of tribe against tribe, to engage all the quarters of the earth in the same work of destruction" (1984, 1039). With ever more effective weapons, destruction as well as production increased. Progress itself was found to be something of a myth where its advocates ignored the complications of high productivity.

In converting what is normally a personified all-governing providence into an impersonal mechanism that brings appropriate rewards to workers, managers, and owners, Adam Smith's invisible hand heads off social turbulence by making "the same distribution of the necessaries of life which would have been made had the earth been divided into equal portions among its inhabitants," *The Theory of Moral Sentiments* (1759) maintains. In the next breath, however, owners and masters collect more benefits than workers because of that same mysterious justice, a first step toward scrapping the last shall be first. Though that contradicts equal portions, so did slavery in belying "all men are created equal." Aloof from imprecation and sacrificial rites, Smith's overseeing power was in accord with deism's impersonal mechanisms. Providence elsewhere sometimes saves and sometimes destroys and does both capriciously as, meanwhile, the economic system distributes its rewards with known hands. Whereas the fuller versions of divine oversight are credited with looking out for crops, weather, and health, the invisible hand works strictly in business affairs.

Except in theocracies, industrial states generally replaced both personified and impersonal oversight with either Smith's lesser mechanism or state managed economy. Thus, to oversimplify as ways to iron out potential rough spots:

Lofty Fictions---->Providence---->Capitalism/Socialist State

One qualification of importance is the continuance of fundamentalism in more than isolated pockets. It indoctrinates each new generation rigorously while other points of view scatter. In Jewish, Christian, and Muslim variants it returns to the fiction of a recently created world, sometimes with an apocalypse to bring fiery destruction followed by eternal peace. The dictionary definition of *ideology* as the science of ideas, abstract speculations, and visionary theorizing fits several other kinds of obsession that resemble fundamentalism. Whereas an ideological state remains on the planet, has no divine overseer, and is scripted only in terms of raw materials, products, and distribution, religious visionaries remove the end state altogether from natural history, which can't be smoothed out and so must be abandoned. It is harsh, beautiful, rough, chaotic, mostly uninhabitable, extreme, and a thousand other things thrown together haphazardly at times and simultaneously rigorous in constants and patterned locally.

Free market ideology is at ease with diverse doctrines so long as they imply that the economy is as it should be even when under Suharto, Pinochet, and Perón. In the heyday of Smith's invisible hand, the robber barons were gathering momentum and slaves were being shipped on one side of a commercial triangle, slaves to the Americas, sugar, cotton, and tobacco to England, textiles, rum, and manufactured goods to Africa. The latter destination was also a source of raw materials. Smith himself, it should be said, was against slavery on both humane and economic grounds, but slaves alone were enough to consign the invisible hand notion to the dustbin. It was as much a visionary dream as the classless state, an ending conflagration, and a new start free of everything the visionary doesn't like. A fair and just distribution of wealth comes from laws and ordinances, not from the natural scheme of things. So do elite hierarchies, egalitarian societies, and tyrannies.

A more critical account of how unregulated economies work is that of William Ashworth in *The Economy of Nature* (1995), where the gains that private exploitation brings are for limited numbers only and for limited times. Lack of effective regulation allows assaults on the environment that can damage it beyond repair, and the losses are permanent for as many as would have benefitted had the soil not eroded, the forest not been stripped, the stream not been polluted, the species not been driven to extinction, the ice caps not melted, and lowlands not flooded. Should deprivation spread to as many as seems likely to climatologists, civil turbulence will spike along with temperatures.

Optimists of the later 18th century were not fully to blame for promoting unlimited resource exploitation, but warning signs were already evident by then. Enough industrial sludge was visible in Smith's day to show the effects of heavy industry on water, air, and land. The romantic poets were aware of them, and Blake (1757-1827) was born just 34 years after Smith (1723-1790). The Marxist version of managed economy is less hospitable to a concept of providence, preferring necessity, in part because Marxism takes over some of the Utmost terminology and the fundamentalist mentality that makes provisional and incomplete information unconditional. It imposes statewide management on an economy that capitalism leaves as unregulated as workers and consumers will allow. Roland Boer in *Criticism of Heaven: On Marxism and Theology* (2007) examines the overlap of religious and economic doctrine in Marxism, concentrating on Ernst Bloch, Walter Benjamin, Louis Althusser, Henri Lefebvre, Antonio Gramsci, Terry Eagleton, Slavoj ŽiŶek, and Theodor Adorno with asides on Bultmann, Max Horkheimer, and a few others. These followers and critics of Marx were aware of the trickiness of capitalist discourse and what it conceals but less aware of the lash-up of Marxism itself. As Robert Nisbet points out in *History of the Idea of Progress* (1980), Marx's protecting workers against alienation from the products of their labor is equivalent to a promised kingdom when the main source of human misery–owners and managers–are replaced (258).

Materializing the terminology of religious final solutions and thereby ending injustice and uncertainty began not with Marx himself but with Kant and Hegel, writing in a language that stirs together theology, ontology, epistemology, and metaphysics. Missing from them and from Marx and Smith is any significant use of natural history in the Bacon, Royal Society, Humboldt, Darwin tradition. It remains largely missing in postmodernism and from pragmatism in James, Dewey, and Rorty (1982, 1991) versions. Marx's commentary on Hegel in *Economic and Philosophic Manuscripts of 1844* (1988) sets aside as irrelevant all aspects of nature outside of production: "*Nature as nature*–that is to say, insofar as it is still sensuously distinguished from that secret sense hidden within it–nature isolated, distinguished from these abstractions, is *nothing*–a nothing *proving itself to be nothing*–is *devoid of sense*, or has only the sense of being an externality which has to be annulled" (166). So Alfred Schmidt (1971) paraphrases with a word tangle similar to that of Hegel and Marx themselves.

In that filing away of natural history, ideology matches eschatology's nature-free zone. Proposing laws of nature is "unthinkable without men's endeavours to master nature," as Schmidt puts it (70). To that way of thinking, economic history begins not with the production of materials, the work of nucleosynthesis, but with their human manipulation. The flaws in that are obvious enough not to need itemizing. Biology is limiting, as hand labor requires hands acquired in the very long and arduous trek from quadrupeds and simians to sapient bipeds. Workers in that capriciously restricted view aren't multifaceted but defined strictly by the jobs they do, as if they, too, didn't exist until they turned up for work. The food they ate for breakfast that powers the hands is prehistorical. So is the plant that went into the oatmeal and the sunlight that powered the photosynthesis. It is as characteristic of ideology as it is of doctrine to declare most of the cosmos off limits. Instead of 'the ways of the Lord are mysterious' as a way to explain something unfortunate, the ideologue focuses on one set of gears in a mechanism. The other gears, belts and chains that drive them, and source of power remain out of sight. So too the brainwork of design and the organizing and management of production that guide the movement of raw materials to consumption. Financing adds bankers and investors to the operation. Making education and the consumer use of products part of the package brings in chefs, housewives, teachers, dishwashers, butchers, and mechanics. Injuries and ailments add doctors, nurses, technicians in labs and imaging, and architects who design hospitals. These all vanish by dictate as if workers weren't born, raised, educated, and trained or suffered handicapping accidents. So do the arts as if novels and plays that shape minds had nothing to do with what those minds then do. The manipulation of ideas is among the more important skill sets, and it isn't limited to the production of goods and services.

Smith and Marx agree on one thing, that the value of commodities stems from the labor put into them, another assumption arbitrary in the manner of ideological myths that lead to the skirmishes of fascism, capitalism, democracy, and communism. It is market value they mean but even there the concept is over simplified. Value attaches to things in demand whatever amount of labor goes into them. Demand leads us into the preferences of minds and the needs of biology, which are initially non economic or pre economic factors. It pertains as much to musicals on Broadway and the Bolshoi Ballet as to shoes and shirts. Natural history as a whole sets the limits and the values, as in the value of oxygen to creatures with lungs. In everything in the natural continuum one thing leads to another. Nothing crops up out of the blue, even the drifting fragments of a chaotic area.

At the very least, supply and demand must be added to the indexing of value, and once that is allowed the value scheme gets complicated. Things laboriously produced but plentiful are valued according to quantity rather than the labor that has gone into them. The value of what appeals to sentiment resides in sensibilities shaped over time, and sentiment is what buys diamonds for engagements, attractive hats to make an impression, and appointments at the hairdresser. Taking it out of marketing would force ad agencies to restructure from the bottom up. Biological needs generate even more demand and go even deeper. Manmade shelters against nature, nature batters down, and so nature creates the demand for more shelters. The value of roofing materials and carpentry is keyed to that. We see something equally extensive in discomforts and hardships that have nothing to do with alienation from the products of labor but with resource limitations and natural disasters, some of them biological. Cancer decreases the value of its victim and increases that of the oncologist and chemotherapy. No utopian state can eliminate any of these things, and capitalism per se is interested in them only insofar as they bring returns to stockholders in the pharmacy industry and makers of medical equipment. Clearly any assessment of commodity value that ignores that range of impingements from nature is excluding vital evidence. Latching onto partial stories as if we were free to disconnect them from the rest of reality moves quickly from tweaking to twisting history.

As I've been arguing, adopting edited stories has an inherent adversarial side. As is clear from its exclusions, Marxism is a comprehensive myth equivalent to beliefs that assign mankind dominion over materials and lesser forms of life. It is inherently antagonistic to monarchies, oligarchies, and capitalist democracies. Both doctrine and ideology extend enticing conclusions to history and start eliminating obstacles ahead of time. Where civil power becomes an extension of an illusion it resorts to Gulags, Lubyanka, and inquisitions. The answers to such departures from reality are always the same, more comprehensive accounts of what can be seen or otherwise detected and is pertinent to an area under scrutiny. What may be the case on other satellites orbiting other stars no one knows, but insofar as nature follows invariables, what holds within the limits of observation holds everywhere. Fish swimming other seas would encounter similar creatures of the deep and perhaps even fishermen. Projecting multiverses with other constants and invariables and possibly one-eyed green creatures remains science fiction. The only universe producing information is the visible one.

Chapter Twelve

Reining in Popular Illusions

Depicting Reality

We have no lack of admirable accounts of how societies and civilizations have formed and organized, mostly as hierarchies with a ruling elite. Elman R. Services's (1975) seminal study *Origins of the State and Civilization* covers much of that multi-cultural history from chiefdoms and egalitarian orders in the "state of nature" to the nearly universal "institutionalization of power" familiar us. Inequality of wealth and power has dominated nearly everywhere since the early dynasties of the Far and Near East and later throughout Europe.

Nor do we lack able defenses of the view that neither modernism nor postmodernism, let alone Parmenides, Zeno, Plato, and others in the neoplatonist tradition, have succeeded in reducing objective nature to shadows or in transferring forms and ideas to an inaccessible realm. That challenges another dominant principle of social orders and ruling hierarchies, their support from imaginary sky powers. The spread of empiricism coincides to some extent with the overthrow of monarchies allied with churches. The serfs and slaves were usually uneducated serfs and slaves. John R. Searle's

(1998) admirably plain style defense of objectivity portrays what nearly everyone finds the case about the objective world and its lack of spirit support for regal offices. The rivers that run through it are truly wet. The mountains standing over them have real trees and rocks. That is a complicated matter much discussed in modern philosophers like John Dewey and Richard Rorty (1982, 1991). Dewey is representative in *Democracy and Evolution* in conceding that although the experience itself is real it has to be formulated "in order to be communicated," and to "formulate requires getting outside of it" (I.2). Once we go from things to discourse about them we encounter difficulties in distinguishing between human inventions and what is truly there.

Overall I've made assumptions similar to Searle's about trusting to a reality perhaps not easily translated into language, math, music, or art but at least free of tricky demons. Searle puts that trust this way: "I regard the basic claim of external realism–that there exists a real world that is totally and absolutely independent of all our representations, all of our thoughts, feelings, opinions, language, discourse, texts, and so on–as so obvious and indeed as such an essential condition of rationality, and even of intelligibility, that I am somewhat embarrassed to have to raise the question and to discuss the various challenges to this view" (14). Except in citing extreme branches of modernism and postmodernism I've also avoided getting tangled in the issues their proponents raise. It should be clear, however, that realism applied to the normal sensory range doesn't make quantum mechanics any less uncertain, make the filtering and translation of objective realism any less problematic, or make a galaxy a million light years away any more accessible. Formulations of this and that and measurements are provisional in both the microcosm and macrocosm and sometimes not all that certain within the visible object range. Neither the age nor the expanse of the universe can be taken as completely settled, though we do know that neither bears any resemblance to the age-old beliefs of every culture or to the myths that popular opinion to this day mistakes for reality. The mechanical sequence set in motion by the inflationary moment and the big bang (or something very much like how they are represented in astrophysics) sent the stars and galaxies forth to cook up the elements.

The odds of an intelligence being on hand to take notice of all this are said to be astronomical without something, somewhere making the right choices. I've cited two cautions against going in that direction. One endorsed by astrophysicists is the equally enormous number of combinations that constants and invariables produce in their mechanical processes. The second consists of two parts, the prominence of inorganic object chaos and the torments of organic life. Nature on its own doesn't think or plan and isn't monstrous. What happens happens. Animated and personified, however, it becomes something altogether different. Its mixture of blessings and living horrors that have been around for hundreds of millions of years become intentional. He, She, or It as an intentional source must like uninhabitable places, relish chaos in the inorganic realm, and in organic nature favor a predatory food chain, and biological ailments, frailty, and mortality. The disintegration phase of the cycles seldom goes peacefully or painlessly. That is why various panaceas have to be established in a separate realm altogether different from the one that could stretch as far as 93 billion light years and contain centillions of integral units.

As to how any similar satellites may have gone or be going through an evolutionary process governed by the same laws, a recent estimate under the direction of Avi Loeb at the Harvard-Smithsonian Center for Astrophysics concludes that life on earth may be premature and anticipate more to come. Another look at the same evidence, however, suggests to others just the opposite, that life here may have already outlived life elsewhere. In either case, life anywhere needs water or some other liquid, no doubt carbon-based chemicals, some source of energy, and a limited range of temperatures in a spectrum from 0 to billions of degrees. Too many stars of about the right size exist not to think that some of them sustain life on one or more satellites, but if any of them do the laws of physics and chemistry guarantee that they follow patterns similar to those within the visible range. The strategy of blaming everything wrong with the planet on mankind was never a feasible alternative even when history was abbreviated to a few thousand years. The irrationality of having lambs devoured by wolves because a man and women were once disobedient isn't one of moral philosophy's better ideas. That for some 3000 years a good many people have taken that guilt as their inheritance is all the demonstration anyone needs for the contagious power of groupthink. The existence of a better place somewhere undiscovered, free of natural laws, would only raise the question, if *there* why not *here,* if *then* why not *now,* and thus throw both moral and natural philosophy back to the age-old question, *why this?* That has a complaint side to it that the natural philosophy question 'why something rather than nothing' doesn't have.

As I've also assumed, of all the ways to limit the troubles grand illusions raise among their many variants, knowledge of natural history and natural philosophy is among the foremost. It is the only authoritative and comprehensive repository of tested theory. Verifiable natural history shuts down some of the pretenses of institutions and dynastic orders. Any belief that is incompatible with it has to be entertained with skepticism. The species that didn't put in an appearance until around the 13.8 billion year mark obviously occupies no privileged position in the real cosmos. Some of the most ardently held convictions could be valid only if the universe were tricky to an incredible degree.

As formidable as getting aligned with reality may seem, it is actually harder to deny what exists and substitute something for it than to acknowledge it. Nature's roughness isn't difficult to see. To deny it requires explaining it away or perhaps writing Philo-like volumes that translate it into messages that only qualified seers and oracles can interpret. The daunting miscellany and the overall dimensions aren't in question. That isn't to say that we have in hand all we might benefit from knowing or ever will have when so many key matters are indeterminable. Not that it is a key matter, but it is impossible to calculate the gravitational influences on any large mass, because they come in the millions and are constantly changing.

Closer to home, we aren't even very certain how our own brains work, they are such a network of connections. The current state of knowledge, including an aspect of human evolution available only to the comparatively new science of epigenetics, the illustrator Brett Ryder (*Johns Hopkins Magazine*, 64.1, spring 2012) compresses into a partitioned image of a brain adorned with a flower, a bird, buildings, terraced farmland, a snail, a honeycomb, a bee, a few geometric shapes, and swirls. Recently revised neuroscience brain maps add hundreds of communicating sectors, some of them working unconsciously. That is all to the good, but the mileage of connecting circuits and the number of neurons and synapses make it impossible to find what goes on physically in the mind at any given moment. Updated diagrams (available on the internet) suggest the magnitude of the problem. Exterior reality is more than a match for that internal circuitry. In both the microcosm and the macrocosm billions have to be multiplied by trillions at times to get the dimensions and relations among very small and very large things.

Fortunately, accuracy down to minuscule fractions isn't necessary to produce highly probable conclusions about a significant amount of data extracted from nature. Going a few digits past the decimal point is good enough for π (pi 3.1415926) as it is for most purposes. To cope with the diversity and massive numbers we *collect, sort* (or classify), *store,* and *transmit* information. The connections among the parts make up a cross-disciplinary network. Arranged as a tree of life, evolution breaks A into A^1 and A^2, and the latter into A^3 and A^4 and so on. Not all variants of all species need be included on a given tree, bush, or shredded fan representation, but if they were, the result could be a tangle looking more like the filaments and patches of gas and dust in galactic superclusters than like a tree or bush.

If that miscellany itself isn't prolific enough to be called chaotic, the connections among the parts are. A great many brilliant workers in the field helped account for as much of it as is now on record, sometimes at peril to themselves from those convinced they would be better served by oracles and myths. At the particle level new discoveries tend to be unsettling, as the possibility of a 5th force operating between electrons and neutrons and ignoring protons is currently doing in some quarters. Particles, hypothetical particles, and anti particles come in undeterminable numbers. As each new piece of equipment goes into operation, however, it brings more under scrutiny. Recently added has been FAST (**F**ive-hundred-meter **A**perture **S**pherical Telescope) in southeast China, the radio astronomy of which can measure cosmic dust, better gauge the spreading rate of the universe, and discover more of the pulsars used for taking measurements of time and space. The map of galaxy clusters and superclusters continues to increase in detail. We can aspire to know much more and not be disappointed.

What we have for planet history and the history of matter can be represented at any length for dissemination and teaching purposes, but for simplicity's sake the animal segment is often shortened to a few icons from a fish coming ashore wobbling on fins to a bent ape and on to an erect mankind. In the history closest to us we've been well served by geology, evolutionary biology, paleontology, and archaeology, and collectively these are more than enough to move ancient and modern misconceptions across the library isles to the make-believe section. In terms of summaries responsible to observable facts outside of book length guides, two convenient reference works are the charts of matter and earth history put together by Paul R. Janke entitled *A Correlated History of Matter* (2004) and *A Correlated History of Earth* (2004) issued by the Worldwide Museum of Natural History. In the possession of more households, these would make excellent myth checking devices. Any such compendium has to cut corners and simplify connecting lines and correlations, but the categories condense volumes of less correlated data into what altogether comes to about the size of two road maps. They earn the correlated title by putting two vertical columns of major eras and periods alongside a master chronology beginning 4.56 billion years ago for the planet and what the chart on matter puts at 13.77 billion years ago. Minerals fill out much of the matter chart, which features the periodic table. From the first violent expansion the universe-to-be underwent an inflationary moment calculated to have ended at 10^{-32} of a second. That fraction, the extended narrative to the present, and an estimated 10^{100+} years to go give a good idea of the spacetime scope into which the story of mankind fits. The magnitude of the power and heat and of the distances, times, and number of stars sets the context for all components.

To cut that very long story short, over 9 billion years after the inflation instant masses of gas and dust and large chunks of heated matter began drawing together under gravity in the solar system. That wasn't a neat and orderly process. Clumps don't occupy the same space gently. Ian Stewart puts it in these terms: "Mathematical models, plus a variety of other evidence from nuclear physics, astrophysics, chemistry, and many other branches of science, have led to the current picture: the planets didn't form as single clumps but by a chaotic process of accretion. For the first 100,000 years, slowly growing 'planetesimals' swept up gas and dust, and created circular rings in the nebula by clearing out gaps between them. Each gap was littered with millions of these tiny bodies." The lumps began "bumping into each other. Some broke up, but others merged; the mergers won and planets built up, piece by tiny piece" (38). The heat from nuclear reactions and the collisions of the planetesimal size masses formed the molten mass that English and German speaking people came to call earth, meaning *ground* or simply *dirt*.

On the Correlated History of Earth chart, lifeforms got underway in prokaryote cells lacking a nucleus and initiated the development of multicellular life. The Cambrian explosion of lifeforms 500 million years ago had over 200 million years of development to go before running into the Permian extinction that eliminated most of them. The now conventional eras and epochs became the Precambrian, Paleozoic, Mesozoic, and Cenozoic periods, established for common use by geologists and paleontologists in time to be available to Charles Lyell (1797-1875). Darwin in the 1840s and 1850s frequently mentions time spans in millions of years that are necessary to explain species changes. The entire natural continuum up to the point of willed behavior and brains capable of making choices depended on environmental conditions. When consciousness began we can't say, but friends of cephalopods like Peter Godfrey-Smith (2017) think that the eye-to-eye contact one can have with them is an exchange of consciousness. Any fish scuttling away from danger or in pursuit of prey has enough of a response system to be called systemic recognition. The overall *correlated* history tells us more about that than any given segment can by itself. It is up to the theory of evolution and natural philosophy to put the pieces together in a developmental line without violating the areas separately. Since many of the species still exist, the study is contemporary as well as historical. It can be both behavioral and biological in examining soft tissue evidence.

The forming of the solar system was part of a chain reaction as everything up to volitional life was. Needless to say all representations selected a limited area. Take a typical contemporary selection with a purposeful design. A road map shows current roadways without indicating why they were placed where they are. A topographical map provides details that influenced the construction without going into underlying factors. A geologic map charts the faults and the movements that shaped the terrain. A plate tectonic map connects the faults to masses moved by molten materials underneath. Water, air, and gravity account for the erosion, the glacial moraine deposits, and the canyons. With that the means of accounting for conditions move into matter prehistory. Convection currents governed by thermodynamics explain the plate movement but not the heat itself. That requires geophysics and another set of graphs and diagrams. A *correlated* history has much of the framework needed to connect a highway map to the rest, but even by itself such a road map is quite an achievement. The brain behind it, the craftsmanship, and use arrived in natural history as a last second development. Nothing in any of the previous eras could manage either the highway or a representation of it.

Scaled representation and symbolic icons put that in teachable form. Mapping separates the parts and puts them in relations based on cause and effect. (*Cause* remember implies nothing about intent in this context.) That is what empiricism attempts to do: be responsive to facts and theory with a high probability factor. Pragmatists like Dewey dedicated to education take what is given. Naturalists add an awareness of the larger context without necessarily citing much of it. Denying the facts doesn't make them go away any more than climate change denial will prevent the seas rising. Each area of the two charts has book-length studies behind it. Anyone wanting more detail will find the following generally readable and informative: P. W. Atkins (1995), Ian Stewart (2016), John Gribbin (2015), Roger Penrose (2004), Sean Carroll (2016), David Eicher (2015), Dave Goldberg and Jeff Blomquist (2010), J Richard Gott (2016), Stephen W. Hawking and Leonard Mlodinow (2010). We've no shortage of such comprehensive works on the Theory of Everything. Any one of these and many another of like kind would have astounded and dismayed students of reality up to the 20th century.

For errors stemming from evidence selection the best correction is to supply what is missing. The 935 lies that the Center for Public Integrity charted in the Bush/Cheney administration's lead-up to the Iraq war were typical in short circuiting intelligence. PolitiFact, a project of the *Tampa Bay Times* taken up by several other newspapers and available online (politifact.com), cites similar examples on a regular non-partisan basis. How secrecy functions in affairs of state is exposed also by inside information leaks. All cultures have a shame component that makes coming out in the open difficult. Explanations of any complexity where justification does exist sound contrived in the reception.

Unfortunately no such devices as PolitiFact or Wikileaks can put natural history into broader circulation. The gap between ruling orders and correlated history is formidable, because not enough voters apply the advanced state of learning to the retarded state of public affairs. The consulting of facts in encyclopedic amounts is a long-term educational project. It has been well underway for decades and received television funding and programming, spectacularly in Carl Sagan's cosmic series resumed by Neil deGrasse Tyson. As I say, the shortcomings lie more in the *application* of the data than in their dissemination. Popular publications such as *Nature, National Geographic,* and the science section of newspapers provide valuable help, but getting across not only the bigger picture but how it applies to reigning beliefs remains an undertaking without a discipline behind it. That isn't to overlook the value of the information itself. In videos good for all ages, for instance, Mr. Parr's internet handling of volcanoes, continental drift, the solar system, and other topics in natural history establishes many of the things it pays to keep in mind. Education at all post grammar school levels has what is needed in some part of the curriculum. What it doesn't usually have is a core course that draws the disciplines together and applies the overall natural history to the rest.

Despite that, not everyone believes even the most attractive myths circulating in their cultures. It couldn't have escaped the attention of everyone in Slavic countries that the star we call the sun didn't need Svarog to make it work. Long before the duration of the universe and of evolutionary biology became known, the Adam and Eve story was illogical enough to raise eyebrows. Very few in India and China gave it serious thought, and to others as well it must have seemed odd that ticks, mosquitoes, precipitous chasms, and cancer in rabbits were owed to a single event in the past. It is easy enough to see the natural phenomena, but drawing logical conclusions from them runs up against custom and the dictates of the subjectivity that comes with individuality.

An Athenian Episode

Let us avoid sensitive recent examples and for a template return to one of intellectual history's more remarkable periods, 5th century Athens. Since we've been there already, the trip this time can skip the scenic sights. Even without significant ideological or doctrinal input mistakes in council chambers and assemblies bring unnecessary suffering in the field. The Melian dialogues and the Cleon/Diodotos debate over the Mytileneans illustrate typical lapses, in those cases less seriously dependent on grand illusions than on suppressed or missing information and on imperialism generally. It was the latter that set the scene and laid the traps. It also helped build the civilization, but I'll skip that oft-told tale. The ethical position of empire-making argued by Pericles was based on his mistaking *common* for *universal* practice in the seizing of the territory and goods of other people. Undisguised, that was armed robbery. Behind its facades it became heroic enterprise. Pericles' unwarranted assertion governs the oft cited passage, "Without being either the ones who made this law [of empire making] or the first to apply it after it was laid down, we applied it as one in existence when we took it up and one that we will leave behind to endure for all time, since we know that you and anyone else who attained power like ours would act accordingly" (5.105). *Anyone* else who attained power is the key. Not everyone but many would be sounder, frequently rather than always. That quantitative distinction makes a considerable difference. Actually most nations most of the time aren't intent on conquering anyone. Some of the best of the world's literature and a strong impetus to undertake inductive investigation came in England before its imperialism took hold. The accomplishments of peace before and after expansion were as evident in Pericles' time as they are now among the nations of the world, most of which stay within their borders.

From the Athenian point of view, the Mytileneans had knowingly broken an alliance. In another framework changing one's alliances to improve the national position is an acceptable practice. When Italy abandoned Germany in 1943, that wasn't considered treachery by anyone except former allies. The US was with the USSR against Germany before it was with Germany against the USSR. In righteous anger the demagogue Cleon puts the case against reversing the judgment of the previous day to kill all adult males and send the rest into slavery (3.36), a judgment some of the Athenians were seriously questioning the next day. The decision hadn't distinguished the innocent from the guilty. Judging from history and from extant tribes, both civilized and primitive societies make that distinction. Omitting evidence of Athenian provocations also distorts the Cleon-backed decision. For those and other reasons, modern readers usually pull for Diodotos to convince the assembly to spare the Mytileneans. He does so in a lengthy argument that underscores the kind of case they are debating and the rules of democratic procedure.

Unfortunately, he also argues on the basis of expediency rather than judiciary principle: "No matter how guilty I proclaim them, I will not on that account urge you to kill them if it is not expedient, nor that because there is some excuse they should keep their city, if it does not appear beneficial [to us]. I consider our deliberations to be more about the future than the present." Or in sum, "We are not taking them to court to get justice but deliberating as to how they might be of use to us" (3.44). *They* again include the innocent as well as the guilty. It would have taken an outsider immune to Athenian ire to point out the motives the Mytileneans found in Athenian aggression or to argue on ethical grounds that punishing everyone for what a few have done is sloppy. Diodotos' point is that realism looks only to consequences. An indiscriminate slaughter of Mytileneans would make other cities at war with Athens fight all the harder knowing what they faced. That the argument is from expediency indirectly excuses the Mytilenean decision to abandon Athens. That has no doubt seemed to them the more prudent course, W. Robert Connor's analysis (1984) of the debate further qualifies the apparent clemency of Diodotos' argument. After the reversed decision, over a thousand Mytileneans were executed. The city was destroyed and the land leased to others whether or not its former owners were guilty of anything. We also know from history that Diodotos is wrong to make passions irresistible and law ineffective as a correction. His pronouncement to that effect is surprising given the legal practices of most states including Athens: it is "very foolish for anyone to believe, that when human nature is eagerly pressing toward some accomplishment, there is some deterrent to stop it by force of law or by any other threat" (3.45). That is another myth, or call it an error of judgment. It too comes under the spell of imperialist ambition and arguments like Pericles' that make the continuation of harsh policies necessary once they are launched. A point of no return is sometimes the case, but many a contention gets settled before the antagonists reach it.

The evidence Diodotos ignores in this case is the moral conditioning citizens of any country undergo from childhood onward. Under their constitution "as it existed before the time of Draco," as Aristotle writes of the source of draconian enforcement, Athenians too supposed that laws have an effect. Polities are normally governable. From monarchy to parliamentary democracy, deterrents are effective in them. They are especially necessary when unscrupulous parties press toward "some accomplishment," which often means conquest. Where bias spreads beyond control and entire nations go berserk, other nations have to step in. Up to that point most of any given citizenry is sensible and humane. Both in this debate and in the slaughter of Melians, many lives hang on the Athenian capacity to reason from evidence, which is neither arcane nor scarce. Their inability to do so and their disregard of justice shows also at Melos, where they have no reason to fear appearing weak. The best place to stifle the unprincipled rule of might is within the offending camp itself. As it is, another atrocity at Milos finishes what misconceptions based on evidence selection have set in motion. The ennobling of the cause added another dimension contributed by the oracles, not surprising in a city named for a goddess. Together these gave free rein to mass slaughter against better instincts that could have been nurtured by a cultivated habit of evidence scrutiny that let some of the air out of self appraisal and national identity. That Thucydides himself is perfectly sound in that regard demonstrates that it could be done.

The Mytilenean incident came late in a series of decisions bearing on Athenian aggression. The Athenians make a similar strained case for it early on in a prewar statement to the Lacedaemonians: "We were compelled from the first by the situation itself to expand the empire to its present state, especially out of fear, then prestige as well, and later out of self-interest. And it no longer seemed safe to risk letting it go when we were detested by most, some had already... been reduced, and you were by then not our friends as you once were but a source of suspicion and contention" (1.75). *Source* equivocates, since the suspicion and contention are mutual and the Athenians have been the more expansive. The alternative? Announcing that expansion was over and another form of consolidation would now replace it if the Spartans agreed could conceivably have encouraged a federation movement rather than a war. Contending parties do meet in water-hole truces, and at that stage no one was at war yet.

Another instance suggests why the Athenians didn't look to their better sources, this one more typical of those misled by lofty myths and unschooled in analytic naturalism. Plutarch (1960) finds the Athenian collapse at Syracuse due in part to science denial, the key figure being the reluctant general Nicias. What he suppressed was the science of lunar eclipses. Anaxagoras had explained how eclipses work, but his explanation hadn't taken hold in popular opinion, which was "instinctively hostile toward natural philosophers and visionaries." Why? Because it "was generally believed that they belittled the power of the gods" (236). That is a familiar refrain among the ancients and remains so among contemporary outposts of religion and superstition. Natural phenomena have to be "communicated with great caution" (236) in any culture that has located itself in a fictitious cosmos.

Because Nicias mistook what a lunar eclipse signified, he followed an outmoded "signs of the gods" superstition and delayed a military move against Syracuse that would probably have succeeded.

In line with that incident and those who preferred oracles to philosophy and science, Protagoras was driven into exile and Anaxagoras imprisoned under distrust. If Plutarch is right, it isn't too much to say that Nicias' superstition cost thousands of Athenian lives, not that as aggressors they deserved a victory that would have resulted in an equal or worse slaughter of Sicilians. Observing humane principles, justice, and common sense would have prevented the armada from sailing in the first place. Proposing not to annex Sicily but to join it to a Greek federation would have been a sensible alternative. Most likely Sicilians would have turned down an invitation, leaving the current situation, but that shouldn't have presented insuperable difficulties to either party. The Athenian war councils also weren't paying attention to Homer, Sophocles, Euripides, Aeschylus, and Aristophanes in whom the foolishness and bitterness of war warned against it. Adding stereotypes to superstition, making rapacity out to be unavoidable, hearkening to oracles, and heeding the ambition of demagogues were more mistakes than the Athenians could overcome.

Evidence Selection

Fact denial comes paired with fact filtering. Minimizing what is unfavorable to a theory is typical of part-for-whole substitutions. In economic matters the part selected may omit what would significantly alter the accounting if it were there, as in these areas of large-scale production and consumption where snap judgment isn't involved but accounting shortcuts are:
- Within a given industry, consumed or lost raw materials such as fish removed from fisheries, soil eroded from farmland, and timber trucked out of forests don't normally figure in

quarterly accounting. That substitution of gain for gain balanced against depletion isn't always necessary, and off the books nearly everyone is aware of it.
- The degradation of rivers, water supplies, and seas from waste, oil spills, and hydrofracking doesn't normally show up on balance sheets, only in government and scientific reports. The non biodegradable trash swirling in the Pacific gyre is an international blight for which no responsibility is assigned. The cost of cleaning up mine tailings and other landlocked materials including nuclear waste gets partly transferred to public funding, as do health costs increased by pollution. These too are unrealistic bookkeeping conventions. Consumers could pay the true cost of what they consume if they appeared in the cost of production and if resources were committed to cleanup.
- The loss of wildlife and biodiversity when industry and populations move or expand is incalculable but real.
- No agency or accounting method covers the full cost of automotive transportation, which includes not only oil spills but injuries from wrecks, air pollution, the building and maintaining of bridges, streets, and highways, the depletion of oil reserves, and the loss of arable land to roadways, services, and suburbs. Environmental degradation, melting polar ice, and climate change are owed fractionally to auto emissions.

Daniel Goleman (1985) adds other deletions, one of them the destruction of Amazon rain forests to produce hamburgers. We know or can find out to the kilowatt how much electricity we have used over a given period. We don't know and probably can't find how many trees went into beef production, let alone the cost of the processing and shipping.

Avoiding mention of realistic costs allows candidates for office and elected officials to put lesser affairs higher on the priority list than more important ones. These are among the inconvenient truths that Al Gore has filed and recited numerous times after hearkening to the testimony of scientists at senate hearings. A major cause of species making the Red List of the IUCN (International Union for Conservation of Nature) is habitat degradation, and there again we usually have no exact accountability, although Goleman suggests that we could calculate "the specific ecological damage involved in a given act of manufacturing" (13) if doing so weren't inconvenient. Since industry officials, lobbyists, and politicians know this, their failure to be forthcoming should probably be classified as dereliction and avoidance rather than ignorance. Voices of caution aren't lacking, but as Dominic Johnson (2004) remarks, positive illusions remain "widespread, pancultural, robust" and are "found in astonishingly diverse contexts" (18). Corrective measures can be broadly philosophic and still have local application. Composed of science and common sense observation, here for instance is an account of natural cycles from Peattie (1941):

> It is even good to die, since death is a natural part of life, a merciful necessity in a world which would only starve and crowd and smother itself if there were no natural mortal end to every existence, no return of the borrowed capital of living matter to the great clearing-house of the mold... The cottontail dashing from the weasel in great, heart-bursting bounds of terror, the turtle sunning himself on a log, the robin turned to let the sunset burn on his red breast while he whistles vespers, all these affirm that life is worth the living. For them their world is a harsh one, despite [the] peace we find in it. The cornered shrew cannot escape the owl's talons; the kinglet dangles on the thorn while the shrike tears it; the woods are not peaceful to those who battle behind the leaves. Yet it is for life that they are fighting; it is to keep the precious gift of it that every creature struggles to the last breath in him (86-87).

That obviously isn't anything for children's library story hour, but not much if anything in it has ever been beyond routine perception. It comes as a surprise to find even Darwin catering to the times in using aesthetic and moral language to describe decidedly unlovely natural phenomena:

> "As natural selection works solely by and for the good of each being, all corporeal and mental endowments will tend to progress towards perfection… There is grandeur in this view of life, with its several powers, having been originally breathed into a few forms or into one; and that, whilst this planet has gone cycling on according to the fixed law of gravity, from so simple a beginning endless forms most beautiful and most wonderful have been, and are being, evolved" (214-215).

More pleasing scenes could be substituted for the one Peattie chooses, but he is balancing the books against such glowing passages as Darwin's, more prominent earlier in the history of ideas than later but still used to support belief in a cosmic order less alien to people. The lessons of naturalism take time to absorb and apply to natural philosophy, and Darwin as a student of Malthus was ordinarily good at doing just that, but on this occasion he takes the 'hardship is good' line of Malthus. That takes a mental effort to screen evidence on a planet so obviously vulnerable to large scale mishap. A forest floor of seedlings is beautiful only when most of the seedlings have died off and left a few to grow to maturity. When a tree falls, decays, and becomes a mother tree to another crowd of seedlings, beauty is again suspended for a time except for those who find decay acceptable, as objectively speaking it should be, being part of a continuous process. The downward spin of a wheel propels the upward spin. The practice of applying aesthetics universally rather than selectively to nature is common also in mathematics, again without the application surveying enough of the field to balance the account. Observable nature on earth's highly productive and rugged surface sets a reasonable standard. The general curvature is geometric and from some angles beautiful. The surface is irregular.

Some of the more capricious gods seem to have been invented to account for such irregularities and perturbations and to suggest by their unpredictability the precarious limbo situation. Georges Roux (1992) reminds us of the realism that went into the creation of the more unpredictable figures. Though he is speaking only of the Tigris-Euphrates valley, the principle holds for other places as well: "The same rivers that bring life can also bring disaster. The winters may be too cold or rainless, the summer winds too dry for the ripening of the dates. A cloudburst can in a moment turn a parched and dusty plain into a sea of mud, and on any fine day a sandstorm can suddenly darken the sky and blow devastation. Confronted with these manifestations of [seemingly] supernatural forces, the Mesopotamian felt bewildered and helpless. He was seized with frightful anxiety" (102). That is the limbo state again when almost anything could happen next, a blessing, a curse, or just another day. Greek mythology reflects a similar realism. To match nature's rough edges takes not only figures of wrath but peevish and capricious ones, or almost as likely in mythology, demons. It was the realistic side of the Hebrews that collected the kindly and cruel sides of former pantheon assortments into a single figure whom they made subject to violent outbursts. When they found the covenant with him not always working, making him a wrathful god was one way to explain why. When the interpreters were priests the national disasters also gave them a chance to blame those who had been derelict in their religious duties.

Profusion and irregularity present formidable obstacles to scientific quantifying. Concerning the body of learning that just his branch of science alone had accumulated by 1941, Peattie acknowledges that after forty years in the field as a professional botanist and amateur ornithologist what he has learned is "just enough to find my way farther." It is "an invitation to limitless discovery" (124). He finds the human intellect overmatched. He could have taken the tone of others and said 'an adventure of discovery', but *invitation* and *limitless* have a more realistic ring to them. Never getting to a decisive conclusion could also be described as getting stuck in limbo, the condition acknowledged by natural philosophy when it discards an impending apocalyptic fury. In looking across the Wyoming plains, Peattie finds a million years minimal for a study of lifeforms. Devoting much less "to the wind in the sage, or to the mysterious urge toward change inherent in all protoplasm" (132) doesn't reveal enough when it comes to evolutionary biology and the bearing that climate and geography have on it. A span of 65 million years would be better, since many former species were eradicated about then, opening the way for mammals. Selecting evidence and spans is sound for most purposes, merely not for the general surveys of natural philosophy.

In a straightforward manner similar to Peatie's, here is Burroughs putting sensations in a context that likewise sets realistic parameters:

> When I go to the woods or fields or ascend to the hilltop, I do not seem to be gazing upon beauty at all, but to be breathing it like the air I would not have the litter and debris removed or the banks trimmed or the ground painted. What I enjoy is commensurate with the earth and sky itself. It clings to the rocks and trees; it is kindred to the roughness and savagery; it rises from every tangle and chasm; it perches on the dry oak stubs with the hawks and buzzards; the crows shed it from their wings and weave it into their nests of coarse sticks; the fox barks it, the cattle low it, and every mountain path leads to its haunts (*America* 54).

Despite the application of *beauty* to litter, debris, roughness, and savagery, such discourse ranks as high as the justice and generosity E. O. Wilson praises in human development. Everyone has the same heritage in that regard. The vast amounts, distances, sizes, and network connections are a lot for myths and illusions to overlook. One normal human reaction in the face of amounts beyond what even supercomputers can handle would be something like a flippant Lewis Carroll saying ("Begin at the beginning and go until you come to the end; then stop"). Or a flippant limerick (Three dozen visionaries sat on peaks/ Counting the visible stars./ When they reached umpteen they ran out of steam/ And settled for Venus and Mars). Or perhaps settle for Fred Hoyle's labeling the incredible, enormous initial singularity the big bang.

References

Albritton, Claude C., Jr. 1980. *The Abyss of Time: Changing Conceptions of the Earth's Antiquity after the Sixteenth Century.* San Francisco: Freeman, Cooper & Co.

Allison, June W. 1989. *Power and Preparedness in Thucydides.* Baltimore: John Hopkins University Press.

Alter, Robert. 1981. *The Art of Biblical Narrative.* New York: HarperCollins.

—. 1985. *The Art of Biblical Poetry.* New York: HarperCollins. Alvis, John. 1995. *Divine Purpose and Heroic Response in Homer and Virgil: The Political Plan of Zeus.* London: Rowman & Littlefield.

Amato, Joseph A. 2000. *Dust: A History of the Small and the Invisible.* Berkeley: University of California Press.

Angier, Natalie. 2006. "Almost Before We Spoke, We Swore," in *The Best American Science and Nature Writing*, ed. Brian Greene. Boston: Houghton Mifflin.

—. 2008. "A Highly Evolved Propensity for Deceit," *New York Times*, December 23.
Apollodorus. 2008. *The Library of Greek Mythology*, trans. Robin Hard. Oxford: Oxford University Press.
Appian, 1996. *The Civil Wars*, John Carter, trans. London: Penguin Books. Aristophanes. 1954. *Four Comedies*. New York: Harcourt, Brace & World. Aristotle. 1951. *The Politics*. New York: Penguin Classics.
Aristotle. 1984. *The Rhetoric and the Poetics*, Edward P. J. Corbett, ed. New York: Modern Library.
Arnhart, Larry. 1998. *Darwinian Natural Right: The Biological Ethics of Human Nature*. Albany: Stare University of New York.
Arnold, Bill T. and Bryan E. Beyer, eds. 2002. *Readings from the Ancient Near East*. Grand Rapids, Michigan. Baker Academic.
Arnold, Thurman W. 1937. *The Folklore of Capitalism*. New Haven: Yale University Press.
Ashliman, D. L. 2004. *Folk and Fairy Tales: A Handbook*. London: Greenwood Press.
Ashworth, William. 1995. *The Economy of Nature: Rethinking the Connections Between Ecology and Economics*. Boston: Houghton Mifflin.
Asimov, Isaac. 1991. *Atom: Journey Across the Subatomic Cosmos*. New York: Dutton.
Atkins, P. W. 1995. *The Periodic Kingdom*. New York: Basic Books.
Atran, Scott. 1990. *Cognitive Foundations of Natural History*. Cambridge: Cambridge University Press.
Atran, Scott and Douglas Medin. 2008. *The Native Mind and the Cultural Construction of Nature*. Cambridge: MIT Press.
Auerbach, Erich. 1953. *Mimesis: The Representation of Reality in Western Literature*. Princeton: Princeton University Press.
Bacon, Francis. 1986 *A Selection of His Works*, ed. Sidney Warhaft. New York: Macmillan.
—. 1996. *Francis Bacon: A Critical Edition of the Major Works*, ed. Brian Vickers. New York: Oxford University Press.
—. 1620, 2009. *The New Organon*, James Spedding, ed. Kindle edition.

—. 1860. *The Works*, James Spedding et al, ed., vol. 4. London: Longman et al.

Bainton, Roland H. 1960. *Christian Attitudes Toward War and Peace.* New York: Abingdon Press.

Bakker, Egbert J. 1997. *Poetry in Speech: Orality and Homeric Discourse.* Ithaca: Cornell University Press.

Baldwin, Charles Sears. 1959. *Medieval Rhetoric and Poetic to 1400.* Gloucester, Massachusetts: Peter Smith.

Barkow, Jerome H. and Leda Cosmides, John Tooby, eds. 1992. *The Adapted Mind: Evolutionary Psychology and the Generation of Culture.* New York: Oxford University Press.

Barnstone, Willis and Marvin Meyer, eds. 2011. *The Gnostic Bible.* Boston: Shambhala.

Barrow, John D. 2000. *The Universe That Discovered Itself.* New York: Oxford University Press.

Barrow, John D. and Frank J. Tipler. 1986. *The Anthropic Cosmological Principle.* New York: Oxford University Press.

Bates, Marston. 1950. *The Nature of Natural History.* Princeton: Princeton University Press.

Beaud, Michel. 2001. *A History of Capitalism: 1500-2000*, trans. Tom Dickman & Anny Lefebvre. New York: Monthly Review Press.

Beckley, Bill, ed. 1998. *Uncontrollable Beauty: Toward a New Aesthetics.* New York: Allworth Press.

Beebee. Thomas O. 1994. *The Ideology of Genre: A Comparative Study of Generic Instability.* University Park: Pennsylvania State University Press.

Berki, R. N. 1977. *The History of Political Thought.* London: Dent.

Berry, William B. N. 1968. *The Growth of a Prehistoric Time Scale.* San Francisco: W. H. Freeman.

Bickerton, Derek. 1990. *Language and Species.* Chicago: University of Chicago Press.

Biddle, Wayne. 1998. *A Field Guide to the Invisible.* New York: Henry Holt and Company.

Bien, Joseph. 1984. *History, Revolution and Human Nature: Marx's Philosophical Anthropology.* Amsterdam: B. R. Gruner

Bierce, Ambrose. 2000. *The Unabridged Devil's Dictionary*, David E. Schultz and S. T. Joshi, eds. Athens, Georgia: University of Georgia Press.

Binford, Lewis R. 1983. *In Pursuit of the Past: Decoding the Archaeological Record.* New York: Thames and Hudson, Inc.

Black, Antony. 2009. *A World History of Ancient Political Thought.* Oxford: Oxford University Press.

Black, Jeremy and Anthony Green. 1992. *Gods, Demons and Symbols of Ancient Mesopotamia.* Austin: University of Texas Press.

Blackmore, Susan. 1999. *The Meme Machine.* Oxford: Oxford University Press.

Bloch, Ernst. 1972. *Atheism in Christianity: The Religion of the Exodus and the Kingdom.* New York: Herder and Herder.

—. 1964, 2000. *The Spirit of Utopia*, trans. Anthony A. Nassar. Stanford: Stanford University Press.

Bloch, Maurice. L 1992. *Prey into Hunter: The Politics of Religious Experience.* Cambridge: Cambridge University Press.

Bloom, Harold. 1990. *The Book of J.* New York: Random House.

—. 2005. *Jesus and Yahweh: The Names Divine.* New York: Riverhead Books.

—. 2000. *Global Brain: The Evolution of Mass Mind from the Big Bang to the 21st Century.* New York: John Wiley.

—. 1995. *The Lucifer Principle: A Scientific Expedition into the Forces of History.* New York: Atlantic Monthly.

Bobbitt, Philip. 2002. *The Shield of Achilles: War, Peace, and the Course of History.* New York: Alfred A. Knopf.

Boehm, Christopher. 1999. *Hierarchy in the Forest: The Evolution of Egalitarian Behavior.* Cambridge: Harvard University Press.

Boer, Roland. 2007. *Criticism of Heaven: On Marxism and Theology* Brill. Boer, Roland. 1997. *Novel Histories: The Fiction of Biblical Criticism.* Sheffield: Sheffield Academic Press.

—. 2009, *Political Myth: On the Use and Abuse of Biblical Themes.* Durham: Duke University Press.

Bowler, Peter J. 2003. *Evolution: The History of an Idea*. Berkeley: University of California Press.

—. 1989. *The Invention of Progress: The Victorians and the Past.* Oxford: Basil Blackwell.

Boyer, Pascal. 2001. *Religion Explained: The Evolutionary Origins of Religious Thought.* New York.: Basic Books.

Bradshaw, Graham. 1993. *Misrepresentations: Shakespeare and the Materialists.* Ithaca: Cornell University Press.

Bragg, Melvyn. 2004. *The Adventure of English.* N.P.: Arcade Publishing.

Braudel, Fernand. 1963, 1995. *A History of Civilizations*, trans. Richard Mayne. London: Penguin Books.

—.1981. *Civilization and Capitalism, 15th-18th Century.* Vol. 1, *The Structures of Everyday Life,* Stan Reynolds, trans. New York: Harper & Row.

Brent, Sandy D. And Ronald L. Geise Jr. 1995. *Cracking Old Testament Codes.* Nashville: B & H Publishing Group.

Brown, Andrew. 1997. *The Darwin Wars.* New York: Simon & Schuster.

Brown, Donald E. 1988. *Hierarchy, History, and Human Nature: The Social Origins of Historical Consciousness.* Tucson: University of Arizona Press.

Brown, Donald E. 1991. *Human Universals.* Philadelphia: Temple University Press.

Brown, Kenneth A. 1993. *Cycles of Rock and Water At the Pacific Edge.* New York: Harper Collins.

Bryson, Bill. 2003. *A Short History of Nearly Everything.* New York: Broadway Books.

Buffon. n.d. *Natural History: Man, the Globe, and of Quadrupeds.* New York: Hurst & Co. Publishers.

Burke, Edmund. 1775, 1939. *Conciliation with the Colonies.* Archibald Freeman and Arthur W. Leonard, eds. Boston: Houghton Mifflin.

Burke, Kenneth. 1941, 1973. *The Philosophy of Literary Form: Studies in Symbolic Action*. Berkeley: University of California Press.

—. 1971. *The Rhetoric of Religion: Studies in Logology*. Berkeley: University of California Press.

Burks, Don M., ed. 1978. *Rhetoric, Philosophy, and Literature*. West Lafayette, Indiana: Purdue University Press.

Burling, Robbins. 2005. *The Talking Ape: How Language Evolved*. Oxford: Oxford University Press.

Burroughs, John. 1977. *America: Selections from the Writings*. Mineola, New York: Dover.

—. 2001. *The Art of Seeing Things*. Syracuse: Syracuse University Press.

Byrne, Richard W. and Andrew Whiten. 1988. *Machiavellian Intelligence: Social Expertise and the Evolution of Intellect in Monkeys, Apes, and Humans*. Oxford: Oxford University Press.

Byrne, Richard. 1995. *The Thinking Ape: Evolutionary Origins of Intelligence*. New York: Oxford University Press.

Carlyle, Thomas, 1971. *Selected Writings*, Alan Shelston, ed. London: Penguin Books.

Carneiro, Robert L. 2003. *Evolution in Cultural Anthropology: A Critical History*. Cambridge, Massachusetts: Westview Press.

Carroll, Joseph. 2004. *Literary Darwinism: Evolution, Human Nature, and Literature*. New York: Routledge.

Carroll, Sean B. 2016. *The Big Picture: On the origins of Life, Meaning and the Universe Itself*. New York: Penguin Random House.

—. 2005. *Endless Forms Most Beautiful: The New Science of Evo Devo*. New York: W. W. Norton.

Carstairs-McCarthy. 1999. *The Origins of Complex Language*. Oxford: Oxford University Press.

Cassirer, Ernst. 1946. *Language and Myth*. New York: Dover.

—. 1946. *The Myth of State*. New Haven: Yale University Press.

Cassirer, Ernst. 1953. *The Philosophy of Symbolic Forms*. 2 vols. New Haven: Yale University Press.

Carter, Rita. 1998. *Mapping the Mind*. Berkeley: University of California Press.
Cawkwell, George. 1997. *Thucydides and the Peloponnesian War*. London: Routledge.
Cherwitz, Richard A., ed. 1990. *Rhetoric and Philosophy*. Hillsdale, J.J.: Lawrence Erlbaum.
Chiapelli, Fredi, ed. 1976. *First Images of America: The Impact of the New World on the Old*. Berkeley: University of California Press.
Childe, V. Gordon. 1951. *Social Evolution*. London: Watts & Co.
Christiansen, Morten H. and Simon Kirby, eds. 2003. *Language Evolution*. Oxford: Oxford University Press.
Chomsky, Noam. 1972. *Language and the Mind*. New York: Harcourt Brace Jovanovich.
—. 1988. *Language and Problems of Knowledge*. Cambridge: MIT Press.
—. 1989. *Necessary Illusions: Thought Control in Democratic Societies*. Boston: South End Press.
—. 1994. *The Prosperous Few and the Restless Many*. Tucson: Odonian Press.
—. 1994. *Secrets, Lies and Democracy*. Tucson: Odonian Press.
—. 1995. *What Uncle Sam Really Wants*. Tucson: Odonian Press.
Chown, Marcus. 2001. *The Magic Furnace: The Search for the Origins of Atoms*. New York: Oxford University Press.
Cicero. 1967. *De Oratore*, 2vols. E. W. Sutton, trans. London: William Heinemann.
Clarke, D. S. 2003. *Sign Levels: Language and Its Evolutionary Antecedents*. Boston: Kluwer Academic Publishers.
Clauswitz, General Carl von. 1874, 2008. *On War*. J. J. Graham, trans. Orlando: Signalman Publishing.
Coffee, Neil. 2009. *The Commerce of War: Exchange and Social Order in Latin Epic*. Chicago: University of Chicago Press.
Collier, Peter. 2014. *A Most Incomprehensible Thing: Notes towards a Very Gentle Introduction to the Mathematics of Relativity*. Kindle Edition.

Conley, Thomas M. 1990. *Rhetoric in the European Tradition*. Chicago: University of Chicago Press.
Connor, W. Robert. 1984. *Thucydides*. Princeton: Princeton University Press.
Coogan, Michael D., ed. (1998). *The Oxford History of the Biblical World*. New York: Oxford University Press.
Cook, Michael. 1986. "The Emergence of Islamic Civilisation," see S. N. Eisenstadt, 476-483.
Cornford, Francis Macdonald. 1907. *Thucydides Mythistoricus*. London: Routledge & Kegan Paul.
Cosmides, Leda and John Tooby. 1994. "Origin of Domain Specificity: The Evolution of Functional Organization," in Hirschfeld and Gelman, *Mapping the Mind*. Cambridge: Cambridge University Press.
Cottrell, Arthur. 2000. *Classical Mythology*. London: Hermes House.
Coward, Rosalind and John Ellis.1977. *Language and Materialism Developments in Semiology and the Theory of the Subject*. London: Routledge and Kegan Paul.
Cremo, Michael A. And Richard L. Thompson. 1994. *The Hidden History of the Human Race*. Badger, Calif.: Govardhan Hill Publishing.
Croix, G.E.M. De Ste. 2004. *Athenian Democratic Origins*. Oxford: Oxford University Press.
Cullman, Oscar. 1964. *Christ and Time: The Primitive Christian Conception of Time and History*. Floyd V. Filson, trans. Philadelphia: Westminster Press.
Dalley, Stephanie, ed. 2000. *Myths from Mesopotamia*. Oxford: Oxford University Press.
Dalrymple, G. Brent. 2004. *Ancient Earth, Ancient Skies: The Age of Earth and Its Cosmic Surroundings*. Stanford: Stanford University Press.
Darwin, Charles. *Charles Darwin on Evolution: the Development of the Theory of Natural Selection*, Thomas F. Glick and David Kohn, eds. Indianapolis: Hackett Publishing, 1996.

—. *The Origin of the Species.* New York: Penguin, 2003.

Davis, Wade. 2002. "The Day the Waylakas Dance," in *When the Wild Comes Leaping Up,* David Suzuki, ed. New York: Greystone Books.

Dawkins, Richard. 2004. *The Ancestor's Tale: A Pilgrimage to the Dawn of Evolution.* Boston: Houghton Mifflin.

—.*The Blind Watchmaker.* London: Penguin Books, 1988. Dawkins, Richard. *A Devil's Chaplain: Reflections on Hope, Lies, Science, and Love.* Boston: Houghton Mifflin, 2003.

—. *The Extended Phenotype: The Long Reach of the Gene.* Oxford: Oxford University Press, 1999.

—. *The God Delusion.* New York: Houghton Mifflin, 2006. Dawkins, Richard, ed. *The Oxford Book of Modern Science Writing.* Oxford: Oxford University Press, 2008.

—. 1998. *Unweaving the Rainbow.* New York: Houghton Mifflin.

Deacon, Terrence W. 1998. *The Symbolic Species.* New York: Norton.

Dean, John W. 2006. *Conservatives Without Conscience.* New York: Viking.

Debnar, Paula, 2004. *Speaking the Same Language: Speech and Audience in Thucydides' Spartan Debates.* Ann Arbor: University of Michigan Press.

De Jong, Irene J. F. 1987. *Narrators and Focalizers: The Presentation of the Story in* The Iliad. Amsterdam: B. R. Grüner.

Dennett, Daniel C. 1991. *Consciousness Explained.* Boston: Little, Brown and Company.

—. 1996. *Darwin's Dangerous Idea: Evolution and the Meanings of Life.* London: Penguin.

Derow, Perter and Robert Parker, eds. 2003. *Herodotus and His World.* Oxford: Oxford University Press.

Derrida, Jaques. 1981. *Positions,* trans. Alan Bass. Chicago: University of Chicago Press.

Descartes, René. 1985. *Discourse on Method,* trans. Laurence J. Lafleur. New York: Macmillan.

—. 1641. 1979. *Meditations on First Philosophy*, trans Donald A. Cress. Indianapolis: Hackett Publishing.

Dever, William G. 2001. *What Did the Biblical Writers Know & When Did They Know It?*. Grand Rapids: William B. Eerdmans Publishing Company.

Dewald, Carolyn J. 2005. *Thucydides' War Narrative*. Berkeley: University of California Press.

Dewey, John. 2015. *The Collected Works*. Kindle edition, including *Creative Intelligence: Essays in the Pragmatic Attitude*, 1917. New York: Henry Holt and Company.

Diamond, Jared. 2005. *Collapse: How Societies Choose to Fall or Succeed*. New York: Viking.

—. 2003. *Guns, Germs, and Steel: The Fates of Human Societies*. New York: W. W. Norton.

—. 1992m 2006. *The Third Chimpanzee: The Evolution and Future of the Human Animal*. New York: Harper Perennial.

Donald, Merlin. 2001. *A Mind So Rare*. New York: Norton.

—. 1991. *Origins of the Modern Mind*. Cambridge: Harvard University Press.

Donne, John. 1966. *John Donne's Poetry*, Arthur L. Clements, ed.. New York: Norton.

Drees, Willem B., ed. 2003. *Is Nature Ever Evil? Religion, Science and Value*. London: Routledge.

Dumont, Louis. 1966, 1980. *Homo Hierarchicus: The Caste System and Its Implications*. Chicago: University of Chicago Press.

Dunbar, Robin, Chris Knight, and Camilla Powers, eds. 1999. *The Evolution of Culture*. New Brunswick, New Jersey: Rutgers University Press.

Dundes, Alan. 1990. *The Hero Pattern and the Life of Jesus* (1976) in *In Quest of the Hero*. Princeton: Princeton University Press.

Durham, William H. 1991. *Coevoution: Genes, Culture, and Human Diversity*. Stanford: Stanford University Press.

Dyson, Julia T. 2001. *King of the Wood: The Sacrificial Victor in Virgil's Aeneid*. Norman, Oklahoma: University of Oklahoma Press.

Eagleton, Terry. 2009. *Reason, Faith, and Revolution: Reflections on the God Debate*. New Haven: Yale University Press.

Earle, Timothy. 2002. *Bronze Age Economics: The Beginnings of Political Economies*. Cambridge, MA.: Perseus Books Group.

—. ed. 1991. *Chiefdoms: Power, Economy, and Ideology*. New York: Cambridge University Press.

Earle, Timothy and Allen W. Johnson. 2000. *The Evolution of Human Societies: From Foraging Group to Agrarian State*. Stanford: Stanford University Press.

Earle, Timothy. 1997. *How Chiefs Come to Power: The Political Economy in Prehistory*. Stanford: Stanford University Press.

Edelman, Gerald M. 1992. *Bright air, Brilliant Fire: On the Matter of the Mind*. New York: Basic Books.

Eicher, David J. 2015. *The New Cosmos: Answering Astronomy's Big Questions*. Cambridge: Cambridge University Press.

Eisenstadt, S. N., ed. 1986. *The Origins and Diversity of Axial Age Civilizations*. New York: State University of New York Press.

Eldredge, Niles. 1982. *The Monkey Business: A Scientist Looks at Creationism*. New York: Washington Square press.

Eliade, Mircea. 1954. 2005. *The Myth of the Eternal Return*. Princeton: Princeton University Press.

—. 1959, 1987. *The Sacred and the Profane: The Nature of Religion*. New York: Harcourt Brace & Company.

Ellis, John M. 1989. *Against Deconstruction*. Princeton: Princeton University Press.

Elster, Jon. 1999. *Alchemies of the Mind: Rationality and the Emotions*. Cambridge: Cambridge University Press.

Emerson, Ralph Waldo. 2000. *The Essential Writings*. New York: Random House.

Englefield, Ronald. 1977. *Language: Its Origin and Relation to Thought*. London: Elek/Pemberton.

Eusebius. 2006. *Ecclesiastical History*, C. F. Cruse, trans. Peabody, Massachusetts: Hendrickson Publishers.

Evelyn, John. 1669, 1968. *The History of the Three Late Famous Impostors*. Los Angeles: Wm.. Andrews Clark Memorial Library.

Falk, Dean. 1992. *Braindance: New Discoveries About Human Origins and Brain Evolution*. New York: Henry Holt.

Ferrill, Arther. 1985. *The Origins of War from the Stone Age to Alexander the Great*. London: Thames and Hudson.

Finkelstein, Israel and Neil Asher Silberman. 2001. *The Bible Unearthed: Archaeology's New Vision of Ancient Israel and The Origin of Its Sacred Texts*. New York: Touchstone.

—. 2006. *David and Solomon: In Search of the Bible's Sacred Kings and the Roots of the Western Tradition*. New York: Free Press.

Finley, M. I. 1985. *The Ancient Economy*. London: Hogarth Press.

Fisch, Harold. 1988. *Poetry with a Purpose: Biblical Poetics and Interpretation*. Bloomington: Indiana University Press.

Fischer, David Hackett. 1970. *Historians' Fallacies*. New York: Harper & Row.

Flandern, Tom Van. 1993. *Dark Matter, Missing Planets, & New Comets*. Berkeley: North Atlantic Books.

Flannery, Tim. 2001. *The Eternal Frontier: An Ecological History of North America and Its Peoples*. New York: Grove Press.

Fodor, Jerry A. 2008. *Lot 2: The Language of Thought Revisited*. Oxford: Oxford University Press.

Folger, Tim. 2005. *The Best American Science and Nature Writing*, ed. Jonathan Weiner. Boston: Houghton Mifflin.

Fortunoff, David. 1993. "Plato's Dialogues as Subversive Activity," in Gerald A Press, ed., *Plato's Dialogues: New Studies & Interpretations*. Lanham, MD: Rowan & Littlefield.

Foucault, Michel. 1972. *The Archaeology of Knowledge*, trans. A. M. Sheridan Smith. New York: Random House.

—. 1970. *The Order of Things: An Archaeology of the Human Sciences*. New York: Random House.

Frankfort, H, H.A. Frankfort, John A. Wilson, and Thorkild Jacobsen. 1949. *Before Philosophy: The Intellectual Adventure of Ancient Man*. Baltimore: Penguin Books.

Frankfort, Henri. 1948. *Kingship and the Gods: A Study of Ancient Near Eastern Religion as the Integration of Society and Nature*. Chicago: University of Chicago Press.

Fried, Morton H. 1967. *The Evolution of Political Society*. New York: Random House.

Friedman, Richard Elliott, ed. 1983. *The Poet and the Historian: Essays in Literary and Historical Biblical Criticism*. Chico, California: Scholars Press.

—. 1987. *Who Wrote the bible*. New York: Summit Books.

Frye, Northrop. 1957. *Anatomy of Criticism*. Princeton: Princeton University Press.

Frye, Northrop and Jay Macpherson. 2004. *Biblical and Classical Myths: The Mythological Framework of Western Culture*. Toronto: University of Toronto Press.

Frye, Northrop. 1982. *The Great Code: The bible and Literature*. New York: Harcourt Brace Jovanovich.

—. 1990. *Words with Power*. New York: Harcourt Brace Jovanovich.

Funkenstein, Amos. 1986. *Theology and the Scientific Imagination from the Middle Ages to the Seventeenth Century*. Princeton: Princeton University Press.

Gabriel, Richard A. and Karen S. Metz. 1991. *From Sumer to Rome: The Military Capabilities of Ancient Armies*. New York: Greenwood Press.

Galbraith, James K. "Exit Strategy: In 1963, JFK Ordered a Complete Withdrawal from Vietnam," *Boston Review*, September 01, 2003.

Gardner, John. 1952. *Fads and Fallacies: In the Name of Science*. New York: Dover.

Gardner, Martin. 1996. *Collected Essays, 1938-1995*. New York: St. Martin's Griffin.

—. 1983, 1999. *The Whys of a Philosophical Scrivener*. New York: St. Martin's Griffin.
Gat, Azar. 2006. *War in Human Civilization*. New York: Oxford University Press.
Goldberg, Dave and Jeff Blomquist. 2010. *A User's Guide to the Universe*. Hoboken, N.J.: John Wiley & Sons.
Gibbon, Edward. 2005. *The History of the Decline and Fall of the Roman Empire*, abridged edition. London: Penguin Books.
Gilbar, Stevens, ed. 1998, *Natural State*: Berkeley: University of California Press.
Gilgamesh. 1999. *The Epic of Gilgamesh*, Andrew George, trans. London: Penguin Books.
Gilovich, Thomas. 1991. *How We Know What Isn't So: The Fallibility of Human Reason in Everyday Life*. New York: The Free Press.
Gilpin, Robert. 1981. *War and Change in World Politics*. Cambridge: Cambridge University Press.
Givon, T. And Bertram F. Malle. 2002. *The Evolution of Language Out of Pre-Language*: Philadelphia: John Benjamins Publishing C.
Gleick, James. 2011. *The Information*. New York: Pantheon Books.
Godfrey-Smith, Peter. 2016. *Other Minds: The Octopus, the Sea, and the Deep Origins of Consciousness*. New York: Farrar, Straus Giroux.
Godfrey-Smith. 2003. *Theory and Reality: An Introduction to the Philosophy of Science*. Chicago: University of Chicago Press.
Goebbels, Joseph. 2009. *The Power of Propaganda*. Kindle edition.
Goffman, Erving. 1959. *The Presentation of Self in Everyday Life*. New York: Random House.
Goleman, Daniel. 1985. *Vital Lies, Simple Truths: The Psychology of Self-Deception*. New York: Touchstone.
Good, Gregory A. 1998. *Sciences of Earth: An Encyclopedia of Events, People, and Phenomena*. New York: Garland Publishing.
Goodman, Nelson. *Languages of Art: An Approach to a Theory of Symbols*. Indianapolis: Bobbs-Merril, 1968.

Gott, J. Richard. 2016. *The Cosmic Web: Mysterious Architecture of the Universe.* Princeton: Princeton University Press.

Gottschall, Jonathan and David Sloan Wilson, eds. 2005. *The Literary Animal: Evolution and the Nature of Narrative.* Evanston: Northwestern University Press.

Gottwald, Norman K. 2001. *The Politics of Ancient Israel.* Louisville, Kentucky: Westminster John Knox Press.

Gould, Stephen Jay. 1996. *Dinosaur in a Haystack.* New York: Crown Trade.

—. 1977. *Ever Since Darwin.* New York: W. W. Norton.

—. 1996. *Full House: The Spread of Excellence from Plato to Darwin.* New York: Three Rivers Press.

—. 2000. *The Lying Stones of Marrakech: Penultimate Reflections in Natural History.* New York: Harmony Books.

—. 2002. *The Structure of Evolutionary Theory.* Cambridge, Massachusetts: Harvard University Press.

—. 1989. *Wonderful Life: The Burgess Shale and the Nature of History.* New York: W. W. Norton.

Graves, Robert. 1992. *The Greek Myths.* London: Penguin Books.

—. 1948. *The White Goddess: A Historical Grammar of Poetic Myth.* New York: Farrar, Straus and Giroux.

Greenberg, Joseph H. 1966, 2005. *Language Universals.* Berlin: Mouton de Gruyter.

Greene, Brian. 1999. *The Elegant Universe: Superstrings, Hidden Dimensions, and the Quest for the Ultimate Theory.* New York: Vintage.

Greene, Kevin. 1983. *Archaeology: An Introduction.* Manchester: Anchor Press.

Grene, David and Richmond Lattimore, eds. 1959. *The Complete Greek Tragedies.* Four volumes, Aeschylus, Sophocles, Euripides. Chicago: University of Chicago Press.

Greenspahn, Frederick E., ed. 1991. *Essential Papers on Israel and the Ancient Near East.* New York: New York University Press.

Greenwood, Emily. 2006. *Thucydides and the Shaping of History*. London: Duckworth.

Gribbin, John. 1999. *Almost Everyone's Guide to Science*. New Haven: Yale University Press.

Gribbin, John and Mary Gribbin. 1996. *Fire on Earth: Doomsday, Dinosaurs, and Humankind*. New York: St. Martin's Press.

Gribbin, John. 2016. *13.8: The Quest to Find the True Age of the Universe and the Theory of Everything*. New Haven: Yale University Press.

Groebel, Jo and Robert A. Hinde, eds.. 1989. *Aggression and War: Their Biological and Social Bases*. Cambridge: Cambridge University Press.

Guilaine, Jean and Jean Zammit. 2005. *The Origins of War: Violence in Prehistory*, trans. Melanie Hersey. Oxford: Blackwell Publishing.

Gunkel, Hermann. 2011. *Israel and Babylon: the Influence of Babylon on the Religion of Israel*. Kindle edition.

Guth, Alan H. 1997. *The Inflationary Universe*. Reading, Massachusetts: Helix Books.

Hagger, Nicholas. 2008. *The Rise and Fall of Civilizations: The Law of History*. Winchester, UK: John Hunt Publishing Ltd.

Hakluyt, Richard. 1598-1600. *The Principal Navigations. Voyages, Traffiques and Discoveries of the English Nation,*.ed. Edmund Goldsmidt. London: G. Bishop, et al.

Hall, Nina, ed. 1991. *Exploring Chaos: A Guide to the New Science of Disorder*. New York: Norton.

Hallpike. C. R. 1979. *The Foundations of Primitive Thought*. Oxford: Clarendon Press.

—. 2008. *How We Got Here: From Bows and Arrows to the Space Age*. Central Milton Keynes: AuthorHouse.

—. 1986. *The Principles of Social Evolution*. Oxford: Clarendon Press.

Halpern, Baruch, 1988. *The First Historians: The Hebrew Bible and History*. San Francisco: Harper & Row.

Hamblin, William J. 2006. *Warfare in the Ancient Near East to 1600 BC: Holy Warriors at the Dawn of History*. London: Routledge.

Harris, Marvin. 2001. *Cultural Materialism*. Walnut Creek, CA.: AltiMira Press.

—. 1999. *Theories of Culture in Postmodern Times*. Walnut Creek, CA.: AltiMira Press.

Harris, Sam. 2005. *The End of Faith: Religion, Terror, and the Future of Reason*. New York: W. W. Norton.

Hart, Donna and Robert W. Sussman. 2005. *Man the Hunted: Primates, Predators, and Human Evolution*. New York: Westview Press.

Hartman, William K & Ron Miller. 1991. *The History of Earth: An Illustrated Chronicle of an Evolving Planet*. New York: Workman Publishing.

Hauser, Marc D. 2000. *Wild Minds: What Animals Really Think*. New York: Henry Holt.

Hawking, Stephen. 1993. *Black Holes and Baby Universes and Other Essays*. New York: Bantam Books.

—. 1996. *A Brief History of Time*. New York: Bantam Books.

Hawking, Stephen and Leonard Mlodinow. 2010. *The Grand Design*. New York: Random House.

Hawking, Stephen. 2001. *The Universe in a Nutshell*. New York: Bantam Books.

Herbert, George. 1953. *The Works*. F. E. Hutchinson, ed. Oxford: Oxford University Press.

Herf, Jeffrey. 2006. *The Jewish Enemy: Nazi propaganda During World Warr II and the Holocaust*. Cambridge: Harvard University Press.

Herodotus. 1972. *The Histories*, trans. Aubrey De Sélincourt. New York: Penguin.

Hesiod. 2004. *Theogony, Works and Days*, and *Shield*, trans. Apostolos N. Athanassakis. Baltimore: Johns Hopkins University Press.

Hesk, Jon. 2000. *Deception and Democracy in Classical Athens*. Cambridge: Cambridge University Press.

Hirschfeld, Lawrence A. and Susan A. Gelman, eds. 1994. *Mapping the Mind: Domain Specificity in Cognition and Culture*. Cambridge: Cambridge University Press.

Hitchens, Christopher. 2007. *God Is Not Great: How Religion Poisons Everything*. New York: Twelve; Hatchette Book Group.

—. 2002. *Unacknowledged Legislation: Writers in the Public Sphere*. London: Verso.

Hitler, Adolf. 1927, 2001. *Mein Kampf*, Ralph Manheim, trans. New York: Houghton Mifflin.

Hitchens, Christopher. 2002. *God Is Not Great: How Religion Poisons Everything*. New York: Twelve; Hatchette Book Group.

—. 2007. *Unacknowledged Legislation: Writers in the Public Sphere*. London: Verso.

Hobbes, Thomas. 1651. *Leviathan*. London: A. Crooke.

—. 1994. *Human Nature* and *De Corpore Politico*, ed J.C.A. Gaskin. Oxford: Oxford University Press.

Hoddeson, Lillian *et al*, eds. 1997. *The Rise of the Standard Model: Particle Physics in the 1960s and 1970s*. Cambridge: Cambridge University Press.

Hoerth, Alfred and John McRay. 2005. *Bible Archaeology: An Exploration of the History and Culture of Early Civilizations*. Grand rapids, Michigan: Baker Books.

Hogan, J. Michael, ed. 1998. *Rhetoric and Community: Studies in Unity and Fragmentation*. Columbia, South Carolina: University of South Carolina Press.

Holstun, James. 2000. *Ehud's Dagger: Class Struggle in the English Revolution*. London: Verso.

—. 1987. *A Rational Millennium: Puritan Utopia of th Seventeenth-Century England and America*. New York: Oxford University Press.

Homer. 1961. *The Iliad*, trans. Richmond Lattimore. Chicago: University of Chicago Press.

Hooke, S. H. 1958. *Myth, Ritual, and Kingship*. Oxford: Oxford University Press.

Horace. 1959. *Satires and Epistles*, trans. Smith Palmer Bovie. Chicago: University of Chicago Press.

Hornblower, Simon, ed. 1994. *Greek Historiography.* Oxford: Oxford University Press.

—. 1987. *Thucydides.* London: Duckworth.

Hoyle, Fred. 1955. *Frontiers of Astronomy.* New York: New American Library.

Hume, David. 2005. *An Enquiry Concerning Human Understanding.* Stilwell, Kansas: Digireads.com.

Humphreys, S. C. 1986. "Dynamics of the Greek Breakthrough: The Dialogue between Philosophy and Religion. See S. N. Eisenstadt, 92- 110.

Jackendoff, Ray. 2002. *Foundations of Language: Brain, Meanng, Grammar, Evolution.* Oxford: Oxford University Press.

—. 1993. *Patterns in the Mind: Language and Human Nature.* New York: Harvester Wheatsheaf.

James, Henry. (2004). *The Portable Henry James.* London: Penguin Books. James, William. 1967. *Essays in Radical Empiricism. A Pluralistic Universe.* Gloucester, Mass.: PeterSmith.

James, William. 2009. *Pragmatism and The Meaning of Truth.* White Dog Publishing.

Jameson, Fredric. 1991, *Postmodernism Or, The Cultural Logic of Late Capitalism.* Durham: Duke University Press.

Jay, Martin. 1984. *Marxism & Totality: The Adventures of a Concept from Lukacs to Habermas.* Berkeley: University of California Press.

Jeans, Sir James. 1932. *The Mysterious Universe.* New York: Macmillan.

Jefferson, Thomas. 1984. *The Writings.* New York: Literary Classics of the United States.

Johanson, Donald and Blake Edgar. 1996. *From Lucy to Language.* New York: Simon & Schuster.

Johansson, Sverker. 2005. *Origins of Language: Constraints on Hypotheses.* Amsterdam: John Benjamins Publishing.

Johnson, Allen W. and Timothy Earle. 2000. *The Evolution of Human Societies: From Foraging Group to Agrarian State*. Stanford: Stanford University Press.

Johnson, Dominic D. P. 2004. *Overconfidence and War*. Cambridge: Harvard University Press.

Jones, Steve. 1999. *Almost Like a Whale: The Origin of Species Updated*. London: Doubleday, 1999.

Josephus. 1998. *The Complete Works*, William Whiston, trans. Nashville: Thomas Nelson.

Josipovici, Gabriel. 1988. *The Book of God*. New Haven: Yale University Press.

—. 1992. *Text and Voice*. New York: St. Martin's Press. Jung, C. G. 1959. *The Archetypes and the Collective Unconscious*, trans. R. F. C. Hull. New York: Pantheon Books.

Kagan, Donald. 1987. *The Fall of the Athenian Empire*. Ithaca: Cornell University Press.

—. 1981. *The Peace of Nicias and the Sicilian Expedition*. Ithaca: Cornell University Press.

Kahneman, Daniel. 2011. *Thinking, Fast and Slow*. New York: Farrar, Straus and Giroux.

Kaku, Michio. 1994. *Hyperspace: A Scientific Odyssey through Parallel Universes*. New York: Oxford University Press.

Kallet, Lisa. 2001. *Money and the Corrosion of Power in Thucydides*. Berkeley: University of California Press.

Kane, Gordon. 2000. *Supersymmetry: Unveiling the Ultimate Laws of Nature*. Cambridge, Massachusetts: Perseus Publishing.

Kang, S-Moon. 1989. *Divine War in the Old Testament and in the Ancient Near East*. Berlin: Walter de Gruyter.

Kant, Immanuel. 1952. *The Critique of Judgement*, trans. James Creed Meredith. Oxford: Oxford University Press.

Kauffman, Stuart. 1995. *At Home in the Universe: A Search for the Laws of Self-Organization and Complexity*. New York: Oxford University Press.

—. 2008. *Reinventing the Sacred: A New View of Science, Reason, and Religion.* New York: Basic Books.
Keeley, Lawrence H. 1996. *War Before Civilization.* New York: Oxford University Press.
Keneally, Thomas. 2007. *Searching for Schindler: A Memoir.* New York: Doubleday.
Kenny, Anthony, ed. 1994. *The Wittgenstein Reader.* Oxford: Blackwell Publishers.
Kingdon, Jonathan. 1993. *Self-Made Man: Human Evolution from Eden to Extinction?* New York: John Wiley & Sons.
Kitchen, K. A. 1966. *Ancient Orient and Old Testament.* Chicago: Inner-varsity Press.
Klein, Naomi. 2007. *The Shock Doctrine: The Rise of Disaste Capitalism.* New York: Henry Holt.
Klein, Richard G. with Blake Edgar. 2002. *The Dawn of Human Culture.* New York: John Wiley & Sons.
Klinkenborg, Verlyn . 2003. *The Rural Life.* Boston: Little, Brown and Company.
Knight, Chris, Michael Studdert-Kennedy, and James R. Hurford, eds. 2000. *The Evolutionary Emergence of Language.* Cambridge: Cambridge University Press.
Knightley, Phillip. 1975. *The First Casualty.* New York: Harcourt Brace Jovanovich.
Koch, Klaus. 1978, 1983. *The Prophets: The Assyrian Period*, trans. Margaret Kohl. Philadelphia: Fortress Press.
—. 1978, 1983, *The Prophets: The Babylonian and Persian Periods.* Philadelphia: Fortress Press.
Kolakowski, Leszek. 1999. *Freedom, Fame, Lying, and Betrayal*, trans. Aghnieszka Kolakowska. Boulder, Colorado: Westview Press.
—. 1978. *Main Currents of Marxism*, vol. 1, trans P:. S. Falla. New York: Oxford University Press.
Koenigsberg, Richard A. 2016. "Hitler as a Political Physician" and "Hitler as the Robert Koch of Germany," Library of Social Science.com

Korten, David C. 1999. *The Post-Corporate World: Life After Capitalism*. San Francisco: Berrett-Koehler Publishers.

Koyré, Alexandre. 1957. *From the Closed World to the Infinite Universe*. Baltimore: Johns Hopkins University Press.

Krause, Sharon R. and Mary Ann McGrail, eds. 2009. *The Arts of Rule*. New York: Rowman & Littlefield Publishers.

Kroll, Richard, Richard Ashcraft, and Perez Zagorin, eds. 1992, *Philosophy, Science, and Religion in England 1640-1700*. Cambridge: Cambridge University Press.

Kuhn, Thomas S. 1996. *The Structure of Scientific Revolution*. 3rd ed. Chicago: University of Chicago Press.

—. 1998. "Objectivity, Value Judgment, and Theory Choice," 435-450 in E. D. Klemke, Robert Holliniger, and David Wyss. 1998. *Philosophy of Science*. Amherst, New York: Prometheus Books.

Kugel, James L. 2003. *The God of Old*. New York: The Free Press.

Kuper, Adam. 1994. *The Chosen Primate: Human Nature and Cultural Diversity*. Cambridge: Harvard University Press.

Lang, Bernhard. 1983. *Monotheism and the Prophetic Minority*. Sheffield: Almond Press.

Lanouette, William, with Bela Silard. 1992. *Genius in the Shadows*. New York: Charles Scribner's Sons.

Larsen, Mogens Trolle, ed. 1979. *Power and Propaganda*. Copenhagen: Akademisk Forlag.

Leakey, Richard. 1994. *The Origin of Humankind*. New York: Harper-Collins.

Lebow, Richard Ned & Barry S. Strauss, eds. 1991. *Hegemonic Rivalry from Thucydides to the Nuclear Age*. San Francisco: Westview Press.

Lederman, Leon with Dick Traversi. 1993. *The God Particle: If the Universe Is the Answer, What is the Question?* New York: Houghton Mifflin.

Leeming, David and Margaret Leeming. 1995. *A Dictionary of Creation Myths*. Oxford: Oxford University Press.

Leeming, David. 2004. *Jealous Gods and Chosen People: The Mythology of the Middle East*. Oxford: Oxford University Press.

—. 2002. *Myth: A Biography of Belief.* New York: Oxford University Press.

—. 2005. *The Oxford Companion to World Mythology.* Oxford: Oxford University Press.

Leopold, Aldo. 1999. *For the Health of the Land.* Washington, D.C.: Island Press.

—. 1949. *Sand County Almanac.* New York: Oxford University Press. Lévi-Srauss, Claude. 1966. *The Savage Mind.* Chicago: Chicago University Press.

—. "The Structural Study of Myth," *Journal of American Folklore*, 68 (1955), 428-444.

Lewis, Cherry. 2000. *The Dating Game: One Man's Search for the Age of the Earth.* Cambridge: Cambridge University Press.

Lieberman, Philip. 1984. *The Biology and Evolution of Language.* Cambridge: Harvard University Press.

—. 1988. *Eve Spoke: Human Language and Human Evolution.* New York: Norton.

Line, R.O.A.M. 1987. *Further Voices in Vergil's* Aeneid. *Oxford: Oxford University Press, 1987.*

Livy. 1971. *The Early History of Rome.* New York: Penguin Books.

Lloyd, G. E. R. 2002. *The Ambitions of Curiosity: Understanding the World in Ancient Greece and China.* Cambridge: Cambridge University Press.

—. 1973. *Greek Science After Aristotle.* New York: W. W. Norton.

—. 1970. *Early Greek Science: Thales to Aristotle.* New York: Norton.

Locke, John. 1690, 2008. *Essay Concerning Human Understanding.* New York: Oxford University Press.

—. 1695, 1997. *The Reasonableness of Christianity.* Washington, D.C.: Regnery Publishing.

—. 1690, 1960. *Two Treatises of Government.* New York: New American Library.

Lockwood, Jeffrey A. 2007. "The Nature of Violence," in *The Best American Science and Nature Writing*, Richard Preston, ed. Boston: Houghton Mifflin.

Logan, Robert K. 2007. *The Extended Mind: The Emergence of Language, the Human Mind, and Culture.* Toronto: University of Toronto Press.

Lorenz, Konrad. 1966, *On Aggression.* New York: Bantam Books.

Losee, John. 1972, 2001. *A Historical Introduction to the Philosophy of Science.* Oxford: Oxford University Press.

Lovejoy, Arthur. 1936, 1964. *The Great Chain of Being.* Luce, T. J. 1997. *The Greek Historians.* London: Routledge.

Lucretius. 1968. *The Way Things Are,* trans. Rolfe Humphries. Bloomington: Indiana University Press.

Luginbill, Robert D. 1999. *Thucydides on War and National Character.* Oxford: Westview Press.

Luraghi, Nino, ed. 2001. *The Historian's Craft in the age of Herodotus.* Oxford: Oxford University Press.

Lyell, Sir Charles. 1866. *Elements of Geology.* New York: Appleton and Company.

Lynch, Aaron. 1996. *Thought Contagion: How Belief Spreads Through Society.* New York: Basic Books, 1996.

Lyon, Thomas J., ed. 1989. *This Incomperable Lande: A Book of American Nature Writing.* New York: Penguin Books, 1989.

Machiavelli, Niccolo. 1961. *The Prince,* George Bull, trans. Baltimore: Penguin Books.

—. 1970. *The Discourses,* ed. Bernard Crick, trans. Leslie J. Walker. New York: Penguin.

McCabe, Joseph. 1985. *History's Greatest Liars.* Austin, Texas: American Atheist Press.

McConnell, Frank, ed. 1986. *The Bible and the Narrative Tradition.* New York: Oxford University Press.

McGhee, George R., Jr. 1996. *The Late Devonian Mass Extinction: The Frasnian/Famennian Crisis.* New York: Columbia University Press.

McKinney, H. Lewis, ed. 1971. *Lamarck to Darwin: Contributions Evolutionary Biology 1809-1859t.o*Lawrence, Kansas:

Coronado Press. Maeterlinck, Maurice. 2006. *The Life of the Bee*. Mineola, New York: Dover.

Maisels, Charles. 2010. *The Archaeology of Politics and Power*. Oxford: Oxbow Books.

——. 1990. *The Emergence of Civilization from Hunting and Gathering to Agriculture, Cities, and the State in the Near East*. London: Routledge. Malthus, Thomas. 1798. *An Essay on the Principle of Population*. London: J. Johnson.

Manchester, William. 1992. *A World Lit Only By Fire: The Medieval Mind and the Renaissance*. New York: Little, Brown and Company.

Markus, George. 1978. *Marxism and Anthropology*. Netherlands: Van Gorcum.

Marvell, Andrew. 1992. *Andrew Marvell*, ed. Frank Kermode and Keith Walker. New York: Oxford University Press.

Marozzi, Justin. 2008. *The Way of Herodotus*. Philadelphia: Da Capo Press. Marx, Karl. 1976. *Capital: A Critique of Political Economy*, trans. Ben Fowkes. Vol 1. London: Penguin.

Marx, Karl. 1844, 1988. *Economic and Philosophic Manuscripts of 1844 and The Communist Manifesto*, trans. Martin Milligan. Amherst, New York: Prometheus Books.

Maeterlinck, Maurice. 1901, 2006. *The Life of the Bee*, trans. Alfred Sutro. Mineola, New York: Dover Publications.

Maynard-Smith, John and Eörs Szathmary. 1999. *The Origins of Life*. Oxford: Oxford University Press.

Mayr, Ernst. 1982. *The Growth of Biological Thought: Diversity, Evolution, and Inheritance*. Cambridge: Harvard University Press.

——. 1964. *Systematics and the Origin of Species*. New York: Dover.

McDougall, Walter A. 1985, 1997. *The Heavens and the Earth: A Political History of the Space Age*. Baltimore: Johns Hopkins University Press.

Merriam-Webster. 1999. *Encyclopedia of World Religions*. Springfield, Mass. Merriam-Webster, Inc.

Merrill, Thomas. 1986. *God-Talk: Paradise Lost and the Grammar of Religious Language*. London: McFarland.

Meyers, Carol. 1988. *Discovering Eve*. New York: Oxford University Press. Miles, Jack. 1995. *God: A Biography*. New York: Vintage Books.

---. 2013. *Rediscovering Eve*. New York: Oxford University Press

Miles, Richard. 2011. *Carthage Must be Destroyed: The Rise and Fall of an Ancient Civilization*. New York: Viking.

Mills, David. 2006. *Atheist Universe: The Thinking Person's Answer to Christian Fundamentalism*. Berkeley: Ulysses Press.

Milton, John. 1957. *Complete Poems and Major Prose*, ed. Merritt Y. Hughes. New York: Odyssey Press.

Mithen, Steven. 1996. *The Prehistory of the Mind: The Cognitive Origins of Art, Religion and Science*. London: Thames and Hudson.

—. 2006. *The Singing Neanderthals: The Origins of Music, Language, Mind, and Body*. Cambridge: Harvard University Press.

Montague, Ashley, ed. 1973. *Man and Aggression*. London: Oxford University Press.

Montaigne, Michel de. 1958. *Essays*, J. M. Cohen, trans. London: Penguin.

Montesquieu, Charles Louis de Secondat. 1750, 1977. *The Spirit of Laws*. Berkeley: University of California Press.

More, Sir Thomas. 1975. *Utopia*, Robert M. Adams, trans. New York: Norton.

Morley, Neville. 2004. *Theories, Models and Concepts in Ancient History*. London: Routledge.

Morris, Simon Conway. 1998. *The Crucible of Creation: The Burgess Shale and the Rise of Animals*. Oxford: Oxford University Press.

Muir, John. 1977. *The Mountains of California*. Berkeley: Ten Speed Press.

Neel, Jasper. 1988. *Plato, Derrida, and Writing*. Carbondale, Ill.: Southern Illinois University Press.

Nelson, Katherine. 1996. *Language in Cognitive Development: Emergence of the Mediated Mind.* Cambridge: Cambridge University Press.

Nemet-Nejat, Karen Rhea. 1998. *Daily Life in Ancient Mesopotamia.* London: Greenwood Press.

Newton, Isaac. 1999. *The Principia: Mathematical Principles of Natural Philosophy*, trans. I. Bernard Cohen and Anne Williams. Berkeley: University of California Press.

Nicolson, Iain. 2007. *Dark Side of the Universe: Dark Matter, Dark Energy and the Fate of the Cosmos.* Johns Hopkins University Press.

Niditch, Susan. 1996. *Oral World and Written Word: Ancient Israelite Literature.* Louisville: Westminster John Knox Press.

—. 1987. *Underdogs and Tricksters: A Prelude to Biblical Folklore.* San Francisco: Harper and Row.

—. 1993. *War in the Hebrew Bible: A Study in the Ethics of Violence.* New York: Oxford University Press.

Nietzsche, Friedrich. 2000. *Basic Writings,* Walter Kaufmann, trans. New York: Random House.

Nisbet, Robert. 1980. *History of the Idea of Progress.* New York: Basic Books.

Norris, Margot. 2000. *Writing War in the Twentieth Century.* Charlottesville: University Press of Virginia.

Not, Martin. 1981. *The Deuteronomistic History.* Sheffield: University of Sheffield Press.

Nye, Bill. 2014. *Undeniable: Evolution and the Science of Creation*, Corey S. Powell, ed. New York: St. Martin's Press.

Ober, Josiah. 1989. *Mass and Elite in Democratic Athens: Rhetoric, Ideology, and the Power of the People.* Princeton: Princeton University Press.

Oldroyd, David R. 1996. *Thinking About the Earth: A History of Ideas in Geology.* Cambridge: Harvard University Press.

O'Malley, John W. 2004. *Four Cultures of the West.* Cambridge, Massachusetts: Harvard University Press.

Orwin, Clifford. 1994. *The Humanity of Thucydides*. Princeton: Princeton University Press.
Osgood, Richard and Sarah Monks, with Judith Toms. 2000. *Bronze Age Warfare*. Gloucestershire: Sutton Publishing.
Osserman, Robert. 1995. *Poetry of the Universe: A Mathematical Exploration of the Cosmos*. New York: Random House.
Otterbein, Keith F. 2009. *The Anthropology of War*. Long Grove, Illinois: Waveland Press.
—. 1994. *Feuding and Warfare*. Langhorne, Penn.: Gordon and Breach.
—. 2004. *How War Began*. College Station: Texas A & M University Press.
Ovid. 1993. *The Metamorphoses*, trans. Allen Mandelbaum. New York: Harcourt.
Pagel, Mark. 2016. "Animal Behaviour: Lethal Violence Deep in the Human Lineage," *Nature*. Published online 28 September 2016.
Pagels, Elaine. 1979. *The Gnostic Gospels*. New York: Random House.
—.1995. *The Origin of Satan*. New York: Random House.
Panofsky, Erwin. 1955. *Meaning in the Visual Arts*. Garden City, New York: Doubleday.
Patai, Raphael. 1990. *The Hebrew Goddess*. Detroit: Wayne State University Press.
Pears, Iain. 2002. *The Dream of Scipio*. New York: Penguin Putnam.
Peattie, Donald Culross. 1935. *Almanac for Moderns*. New York: Putnam's Sons.
Peattie, Donald Culross. 1938. *A Prairie Grove: A Naturalist's Story of Primeval America*. New York: The Literary Guild.
—. 1941. *The Road of a Naturalist*. New York: Houghton Mifflin.
Pelling, Christopher. 2002. *Plutarch and History: Eighteen Studies*. London: Gerald Duckworth.
Pendergrast, Mark. 2003. *Mirror \ Mirror : A History of the Human Love Affair with Reflection*. New York: Basic Books.

Penrose, Roger. 1989. *The Emperor's New Mind: Concerning Computers, Minds, and the Laws of Physics*. Oxford: Oxford University Press.

—. 2005. *The Road to Reality: A Complete Guide to the Laws of the Universe*. New York: Random House.

—. 1994. *Shadows of the Mind: A Search for the Missing Science of Consciousness*. Oxford: Oxford University Press.

Peters, F. E. 1982. *Children of Abraham: Judaism/Christianity/Islam*. Princeton: Princeton University Press.

—. 2003. *Judaism, Christianity and Islam: The Monotheists*. New York: Recorded Books, LLC.

Phillips, Kevin. 1990. *The Politics of Rich and Poor*. New York: HarperCollins.

Philo. 1993. *The Works of*, C. D. Yonge, trans. Peabody, Massachusetts.

Pigliucci, Massimo. 2002. *Denying Evolution: Creationism, Scientism, and the Nature of Science*. Sunderland, Massachusetts: Sinauer Associates, Publishers.

Pinker, Steven. 1995. *The Language Instinct*. New York: Harper Perennial.

Plato. 1961. *The Collected Dialogues*. Princeton: Princeton University Press. Plotkin, Henry. 1994. *The Nature of Knowledge*. London: Penguin Books. Pluciennik, Mark. 2005. *Social Evolution*. London: Gerald Duckworth.

Plutarch. 1894, 2009. *The Lives*, vols 1-3, trans. Aubrey Stewart and George Long. London: George Bell & Sons.

—. 1960. *The Rise and Fall of Athens: Nine Greek Lives*, Ian Scott-Kilvert trans. London: Penguin Books.

Pollack, Robert. 1999. *The Missing Moment: How the Unconscious Shapes Modern Science*. Boston: Houghton Mifflin.

Pollan, Michael. 2006. *The Omnivore's Dilemma: A Natural History of Four Meals*. New York: Penguin Books.

Polybius. 1922-1927. *The Histories*. Loeb Classical Library (website) Edition.

—. 1966. *The Histories*, trans. Mortimer Chambers. New York: Washington Square Press.

Ponting, Clive. 1991. *A Green History of the World*. New York: Penguin.

Popkin, Richard H. 1980. "Jewish Messianism and Christian Millenarianism," in Perez Zagorin, ed. *Culture and Politics from Puritanism to the Enlightenment*. Berkeley: University of California Press.

Popper, Karl. 1935, 2002. *The Logic of Scientific Discovery*. London: Routledge.

Powell, James Lawrence. 2001. *Mysteries of Terra Firma: The Age and Evolution of the World*. New York: Free Press.

—. 1998. *Night Comes to the Cretaceous*. New York: Harcourt Brace & Company.

Power, Margaret. 1991. *The Egalitarians–Human and Chimpanzee*. Cambridge: Cambridge University Press.

Preston, Richard, ed. 2007. *The Best American Science and Nature Writing*. New York: Houghton Mifflin.

Pritchard, James B., ed. 1969. *The Ancient Near East: Supplementary Texts and Pictures Relating to the Old Testament*. Princeton: Princeton University Press.

—. ed. 1955. *Ancient Near Eastern Texts Relating to the Old Testament*. Princeton: Princeton University Press.

Puttenham, George. 1589. *The Arte of English Poesie*. London: Richard Field.

Quine, W. V. 1992. *Pursuit of Truth*. Cambridge: Harvard University Press.

Quintilian. 1958. *The Institutio Oratoria*, H. E. Butler, trans. Cambridge: Harvard University Press.

Raaflaub, Kurt A., ed. 2007. *War and Peace in the Ancient World*. Oxford: Blackwell Publishing.

Raff, Rudolf. A. 1996. *The Shape of Life: Genes, Development, and the Evolution of Animal Form*. Chicago: University of Chicago Press.

Raglan, Lord. 1990. *The Hero: A Study in Tradition, Myth, and Drama* part III (1956) in *In Quest of the Hero*. Princeton: Princeton University Press.

Rank, Otto. 1990. *The Myth of the Birth of the Hero* (1959, 1983) in *In Quest of the Hero*. Princeton: Princeton University Press.

Redman, Charles L. 1978. *The Rise of Civilization: From Early Farmers to Urban Society in the Ancient Near East*. San Francisco: W. H. Freeman and Company.

Reed, Walter L. 1993. *Dialogues of the Word: The Bible as Literature According to Bakhtin*. New York: Oxford University Press.

Rees, Martin. 2000. *Just Six Numbers*. New York: Basic Books.

Rhodes, P. J., ed. 2004. *Athenian Democracy*. Edinburgh: Edinburgh University Press.

Richards, Jennifer. 2008. *Rhetoric*. London: Routledge.

Richter, Horst-Eberhard. 1984. *All Mighty: A Study of the God Complex in Western Man*, trans. Jan van Heurck. Claremont, California: Hunter House.

Ridley, B. K. (1984). *Time, Space and Things*. Cambridge: Cambridge University Press.

Ridley, Matt. 1999. *Genome: The Autobiography of a Species in 23 Chapters*. New York: Perennial.

—. 1996. *The Origins of Virtue: Human Instincts and the Evolution of Cooperation*. New York: Penguin Books.

Rindos, David.; 1984. *The Origins of Agriculture:An Evolutionary Perspective*. New York: Harcourt Brace Jovanovich.

Roaf, Michael. 1990. *Cultural Atlas of Mesopotamia and the Ancient Near East:* Oxford: Equinox.

Roberts, Kenneth. 1930. *Arundel*. Camden, Maine: Down East Books.

Robertson, David. 1977. *The Old Testament and the Literary Critic*. Philadelphia: Fortress Press.

Rorty, Richard. 1982. *Consequences of Pragmatism (Essays: 1972-1980)*. Minneapolis: University of Minnesota Press.

—. 1991. *Objectivity, Relativism, and Truth*. Cambridge: Cambridge University Press.

Rosenberg, Joel. 1986. *King and Kin: Political Allegory in the Hebrew Bible*. Bloomington: Indiana University Press.
Rosmarin, Adena. 1985. *The Power of Genre*. Minneapolis: University of Minnesota Press.
Ross, Don, Andrew Brook, and David Thompson, eds. 2000. *Dennett's Philosophy: A Comprehensive Assessment*. Cambridge: MIT Press.
Ross, Eric B., ed. 1980. *Beyond the Myths of Culture: Essays in Cultural Materialism*. New York: Academic Press.
Roth, Martha T. 1998. "Gender and Law: A Case Study from Ancient Mesopotamia," in Victor H. Matthews, et al, eds, *Gender and Law in the Hebrew Bible and the Ancient Near East*. Sheffield: Sheffield Academic Press.
Rousseau, Jean-Jacques. 1997. *The Discourses and Other Early Political Writings*, Victor Gourevitch, ed. Cambridge: Cambridge University Press.
Roux, Georges. 1966. *Ancient Iraq*. London: Penguin Books.
Rudgley, Richard. 1999, *The Lost Civilizations of the Stone Age*. New York: Free Press.
Rudwick, Martin J. S. 1985. *The Meaning of Fossils: Episodes in the History of Paleontology*. Chicago: University of Chicago Press.
Ruether, Rosemary Radford. 2005. *Goddesses and the Divine Feminine*. Berkeley: University of California Press.
Ruse, Michael. 2004. *Can a Darwinian Be a Christian?* Cambridge: Cambridge University Press.
—. 2003. *Darwin and Design: Does Evolution Have a Purpose?* Cambridge: Harvard University Press.
Ruskin, John. 2006. *The Ethics of the Dust*. N.P.: Bibliobazaar.
Russell, Bertrand. 1935, 1997. *Religion and Science*. New York: Oxford University Press.
Ryle, Gilbert. 1949. *The Concept of Mind*. Chicago: University of Chicago Press.
Sacks, Kenneth. 1981. *Polybius on the Writing of History*. Berkeley: University of California Press.

Sagan, Carl. 1980. *Cosmos*. New York: Wings Books.

—. 1996. *The Demon-Haunted World: Science As a Candle in the Dark*. New York: Ballantine Books.

—. 1994. *Pale Blue Dot: A Vision of the Human Future in Space*. New York: Random House.

Salinger, Pierre and Leonard Gross, 1988. *Mortal Games*. New York: St. Martin's Press.

Sandby, D. Bent and Ronald L. Giese, Jr. 1995. *Cracking Old Testament Codes: a Guide to Interpreting the Literary Genres of the Old Testament.*. Nashville: Broadman and Holman, Publishers

Sanderson, Stephen K. 2001. *The Evolution of Human Sociality: A Dawinian Conflict Perspective*. New York: Rowman and Littlefield.

—. 1999. *Social Transformations: A General Theory of Historical Development*. New York: Rowman and Littlefield.

Saner, Reg. 2010. *Living Large in Nature: A Writer's Idea of Creationism*. Chicago: The Center for American Places at Columbia College Chicago.

Sarles, Harvey B. 1977. *Language and Human Nature*. Minneapolis: University of Minnesota Press.

Scardigli, Barbara, ed. 1995. *Essays on Plutarch's* Lives. Oxford: Oxford University Press.

Scaruffi, Piero. 2009. *Wars and Genocides of the 20th Century*. Scaruffi.com/politics/massacre.

Schama, Simon. 1995. *Landscape and Memory*. New York: Alfred A. Knopf.

Schmidt, Alfred. 1962, 1971. *The Concept of Nature in Marx*, trans. Ben Fowkes. London: NLB..

Schwartz, Jeffrey H. 1999. *Sudden Origins: Fossils, Genes, and the Emergence of Species*. New York: John Wiley and Sons. Shakespeare,

William. 1992. *The Complete Works*, David Bevington, ed. New York: HarperCollins.

Sharp, Carolyn J. 2009. *Irony and Meaning in the Hebrew Bible*. Bloomington: Indiana University Press.

Shattuck, Roger. 1996. *Forbidden Knowledge: From Prometheus to Pornography*. New York: St. Martin's Press.

Shennan, Stephen. 2002. *Genes, Memes and Human History: Darwinian Archaeology and Cultural Evolution*. London: Thames & Hudson.

Shermer, Michael. 2011. *The Believing Brain: From Ghosts and Gods to Politics and Conspiracies*. New York: Henry Holt and Company.

Sidney, Sir Philip. 1992. *Sir Philip Sidney*, Katherine Duncan-Jones, ed. New York: Oxford University Press.

Singer, Peter. 1999. *A Darwinian Left: Politics, Evolution and Cooperation*. London: Weidenfield & Nicolson.

Smith, Adam. 1985. *Lectures on Rhetoric and Belles Lettres*. Indianapolis: Liberty Fund.

—. 1759. *The Theory of Moral Sentiments*. Edenburgh: A. Kincaid and J. Bell.

—. 2003. *The Wealth of Nations*. New York: Random House. Smith, Cameron M. and Charles Sullivan. 2007. *The Top 10 Myths about Evolution*. Amherst, New York: Prometheus Books.

Smith, Mark. S. 1990, 2002. *The Early History of God: Yahweh and the Other Deities in Ancient Israel*. Grand Rapid, Michigan: Wm. B. Eerdmans Publishing Co.

—. 2004. *The Memoirs of God: History, Memory, and the Experience of the Divine in Ancient Israel*. Minneapolis: Fortress Press.

Smolin, Lee. 2007. *The Trouble with Physics*. Boston: Houghton Mifflin.

Sorabji, Richard and David Rodin, eds. 2006. *The Ethics of War*. Burlington, VT: Ashgate Publishing Co.

Soukhanov, Anne H. 1995. *Word Watch: The Stories Behind the words of Our Lives*. New York: Henry Holt and Company.

Sparks, Kenton L. 2005. *Ancient Texts for the Study of the Hebrew Bible*. Peabody, Mass.: Hendrickson Publishers.

Spengler, Oswald. 1926, 1928. *The Decline of the West*, 2 vols. Trans. Charles Francis Atkinson. New York: Alfred A. Knopf.

Spinoza, Benedict. 1670. *A Theologico-Political Treatise (Theologico-Politicus)*, trans. R. H. M. Elwes. MobileReference.

Stadter, Philip A, ed.. 1992. *Plutarch and the Historical Tradition*. London: Routledge.

Stark, Rodney. 1997. *The Rise of Christianity*. San Francisco: Harper.

Starke, Linda, ed. *The State of the World, 2006*. New York: W. W. Norton & Company.

Statius, Publius Papinius. 2004. *The Thebaid: Seven against Thebes*, Charles Stanley Ross, trans. Baltimore: Johns Hopkins University Press.

Stegner, Wallace and Page Stegner. 1981. *American Places*. New York: E.P. Dutton

Stegner, Wallace. 1962, 1990. *Wolf Willow*. New York: Penguin Books.

Steinhardt, Paul J. and Neil Turok. 2007. *Endless Universe: Beyond the Big Bang*. New York: Random House.

Stern, Menahem, ed. 1974. *Greek and Latin Authors on Jews and Judaism*, vol. 1. Jerusalem: Israel Academy of Science and Humanities.

Sternberg, Meir. 1987. *The Poetics of Biblical Narrative: Ideological Literature and the Drama of Reading*. Bloomington: Indiana University Press.

Stevens, Wallace. 1966. *The Letters of*, ed. Holly Stevens. New York: Alfred A. Knopf.

—. 1942. *The Necessary Angel: Essays on Reality and the Imagination*. New York: Alfred A. Knopf.

—. 1971. *The Palm at the End of the Mind*, ed Holly Stevens. New York: Alfred A. Knopf.

Stewart, Ian. 2016. *Calculating the Cosmos: How Mathematics Unveils the Universe*. London: Profile Books.

Stone, Michael E. 2011. *Ancient Judaism*. Grand Rapids: Eerdmans Publishing.

Storey, Robert. 1996. *Mimesis and the Human Animal: On the Biogenetic Foundations of Literary Representation*. Evanston: Northwestern University Press.

Susskind, Leonard. 2006. *The Cosmic Landscape: String Theory and the Illusion of Intelligent Design*. New York: Little, Brown and Company.

Suzuki, David, ed. 2002. *When the Wild Comes Leaping Up: Personal Encounters with Nature*. New York: Greystone Books.

Sverker, Johansson. 2005. *Origins of Language: Constraints in Hypothesis*. Amsterdam: John Benjamins Publishing Company.

Sweeney, Marvin A. and Ehud Ben Zvi, eds. 2003. *The Changing Face of Form Criticism*. Grand Rapids: Wm. B. Eerdmans Publishing.

Sykes, Bryan. 2001. *The Seven Daughters of Eve*. New York: Norton & Company.

Tacitus. 1997. *The Histories*, trans. W. H. Fyfe, revised and edited by D. S. Levene. Oxford: Oxford University Press.

Tallerman, Maggie, ed. 2005. *Language Origins: Perspectives on Evolution*. Oxford: Oxford University Press.

Tattersall, Ian. 1998. *Becoming Human: Evolution and Human Uniqueness*. New York: Harcourt Brace.

Tavris, Carol and Elliot Aronson. 2007. *Mistakes Were Made (but not by me)*. New York: Harcourt.

Thomas, Lewis. 1974. *The Lives of a Cell*. New York: Penguin.

Thomas, Rosalind. 2000. *Herodotus in Context: Ethnography, Science and the Art of Persuasion*. Cambridge: Cambridge University Press.

Thompson, Thomas L. 1999. *The Mythic Past: Biblical Archaeology and the Myth of Israel*. London: Random House.

Thoreau, Henry David. 1960. *Walden and "Civil Disobedience*. New York: Signet.

Thucydides. 1998. *The Peloponnesian War*, trans. Steven Lattimore. Indianapolis: Hackett Publishing Company.

De Tocqueville, Alexis. 2003. *Democracy in America and Two Essays on America*, Gerald E. Bevan, trans. London: Penguin Books.

Toliver, Harold. 1974. *Animate Illusions: Explorations in Narrative Structure*. Lincoln: University of Nebraska Press.

—. 2009. "Numbered Days," 28-31. *Oregon Quarterly*, volume 88: 4

—. 1981, *The Past That Poets Make*. Cambridge: Harvard University Press.

Tomasello, Michael. 1999. *The Cultural Origins of Human Cognition*. Cambridge: Harvard University Press.

Trefil, James. 1997. *Are We Unique?* New York: John Wiley & Sons.

—. 2003. *The Nature of Science*. New York: Houghton Mifflin.

Trigger, Bruce G. 1989. *A History of Archaeological Thought*. Cambridge: Cambridge University Press.

—. 2003. *Understanding Early Civilizations*. Cambridge: Cambridge University Press.

Tudge, Colin. 1999. *Neanderthals, Bandits, and Farmers: How Agriculture Really Began*. New Haven: Yale University Press.

—. 1997. *The Time Before History: 5 Million Years of Human Impact*. New York: Touchstone.

Turner, Frederick Jackson. 1920, 1996. *The Frontier in American History*. New York: Dover Publications.

Turney-High, Harry Holbert. 1949. *Primitive War*. Columbia, South Carolina: University of South Carolina Press.

Tyson, Neil deGrasse, Michael A. Strauss, and J. Richard Gott. 2016. *Welcome to the Universe*. Princeton: Princeton University Press.

Uffenheimer, Benjamin. 1986. "Myth and Reality in Ancient Israel." See S. N. Eisenstadt, 135-168.

Vaihinger, Hans. 1935. *The Philosophy of 'As If': A System of Theoretical, Practical and Religious Fictions of Mankind*, trans. C. K. Ogden. London: Routlede and Kegan Paul.

Van Flandern, Tom. 1993. *Dark Matter, Missing Planets and New Comets*. Berkeley: North Atlantic Books.

Van Setters, John. 1983. *In Search of History: Historiography in the Ancient World and the Origins of Biblical History*. New Haven: Yale University Press.

Vaughan, Henry. 1650. 1964. *The Complete Poetry,* French Fogle, ed. New York: Norton.

Vayda, Andrew P. 1976. *War in Ecological Perspective.* New York: Plenum Press.

Veblen, Thorstein. 1899, 1958. *The Portable Veblen.* New York: Viking Press.

Versluis, Arthur. 2001. *Way of Native American Traditions.* London: Thorsons.

Vickers, Brian, ed. 1987. *English Science, Bacon to Newton.* Cambridge: Cambridge University Press.

—. 1988. *In Defense of Rhetoric.* Oxford: Oxford University Press.

Virgil. 1964. *Aeneid,* Books I-VI, ed. Clyde Pharr. Wauconda, Illinois: Bolchazy-Carducci Publishers.

—. 1990. *The Aeneid,* trans. Robert Fitzgerald. New York: Random House.

Walbank, Frank W. 2002. *Polybius, Rome and the Hellenistic World: Essays and Reflections.* Cambridge: Cambridge University Press.

de Waal, Frans. 2005. *Our Inner Ape.* New York: Penguin.

Wallerstein, Immanuel. 2000. *The Essential Wallerstein.* New York: The New Press.

Wegener, Alfred. 1928, 1966. *The Origin of Continents and Oceans,* trans. John Biram. New York: Dover Publications.

Weinfeld, Mosche. 1 986. "The Protest against Imperialism in Ancient Israelite Prophecy." See S.. N. Eisenstadt, 169-182.

Weisman, Alan. 2007. *The World Without Us.* New York: St. Martin's Press.

Wellhausen, Julius, 1957. *Prolegomena to the History of Ancient Israel.* New York: Meridian Books.

Wells, H. G. 1991 (1922). *A Short History of the World.* New York: Penguin.

Wells, Spencer. 2002. *The Journey of Man: A Genetic Odyssey.* New York: Random House.

Weatherall, James Owen. 2016. *Void: The Strange Physics of Nothing.* New Haven: Yale University Press.

Wills, Lawrence M. 1995. *The Jewish Novel in the Ancient World.* Ithaca: Cornell University Press.
Williams, George C. 1966. *Adaptation and Natural Selection.* Princeton: Princeton University Press.
Wilson, David Sloan. 2002. *Darwin's Cathedral: Evolution, Religion, and the Nature of Society.* Chicago: University of Chicago Press.
Wilson, Edward O. 1998. *Consilience: The Unity of Knowledge.* New York:AlfredA.Knopf.
—. 1999. *The Diversity of Life.* New York: W. W. Norton.
—. 1978. *On Human Nature.* Cambridge: Harvard University Press.
Wilson, Robert R.1980. *Prophecy and Society in Ancient Israel.* Philadelphia: Fortress Press.
Wilson, Thomas. 1574, 1982. *Arte of Rhetorique,* Thomas J. Derrick, ed. New York: Garland.
Wiseman, T. P. 1994. *Historiography and Imagination.* Exeter: University of Exeter Press.
Wootton, David, ed. 1986. *Divine Right and Democracy: An Anthology of Political Writing in Stuart England.* London: Penguin.
Wrangham, Richard and Dale Peterson. 1996. *Demonic Males: Apes and the Origins of Human Violence.* Boston: Houghton Mifflin.
Hamilton, James Wright, Benjamin F., ed. 1961, 2002.
The Federalist, by Alexander Hamilton, Madison, and John Jay. New York: MetroBooks.
Xenophon. 2010. *Cyropaedia: The Education of Cyrus,* Henry Graham Dakyns, trans. Kindle edition.
—. 1891, 1998. *Hellenica,* H. G. Dakyns, trans. London: . Cassell & Company.
—. 1888. *The Memorable Thoughts of Socrates,* Edward Byshe, trans. Digital.
—. 1972. *The Persian Expedition.* London: Penguin.
Yeats, W. B. 1990. *The Collected Works of W. B. Yeats.* Vol 1, *The Poems.* New York: Macmillan.
Zagorin, Perez. 1998. *Francis Bacon.* Princeton: Princeton University Press.

———. 1990. *Ways of Lying: Dissimulation, Persecution, and Conformity in Early Modern Europe.* Cambridge: Harvard University Press.

Zimbardo, Philip. 2007. *The Lucifer Effect: Understanding How Good People Turn Evil.* New York: Random House

Zimmer, Carl. 2006. "Devious Butterflies, Full-Throated Frogs and Other Liars," *New York Times,* December 26.

Index

A

Aeschylus 49, 223, 315
Alexander 207, 223, 249, 262
algorithm 47, 61, 123, 156
allegory 115, 123, 127, 229
Alter, Robert 163
American Civil War 191, 279, 280, 281
Anaxagoras 314, 315
ancient Egypt xii, 89, 108, 122, 126, 153, 156, 159, 198, 203, 205, 206, 220, 221, 222, 230, 232, 244, 247, 285
ancient Greece. *See* Aristotle, Athens, Hesiod, Homer, Plato, Thucydides ancient Israel; *See also* Hebrew anthology
ancient Mesopotamian dynasties xv, 89, 91, 111, 117, 122, 125, 126, 134, 141, 146, 150, 151, 152, 156, 157, 198, 203, 206, 220, 221, 230, 246, 252, 319
ancient Persia xv, 49, 122, 125, 202, 203, 208, 221, 222, 225, 232, 235, 247. *See also* Xerxes ancient Rome; *See* Lucan, Lucretius, Ovid, Plutarch, Tacitus, Virgil
Anderson, Benedict 117, 249
Aquinas, Thomas 61, 175
archaeology xv, xviii, 124, 147, 214, 229, 293, 306
Aristophanes 238, 315
Aristotle 9, 21, 61, 107, 108, 114, 115, 116, 123, 124, 125, 134, 175, 181, 313
Arnold, Bill T. 151, 153, 163
Ashliman, D. L. 116
Ashworth, William 295
as if 74, 233, 242
Athens xviii, 49, 102, 124, 222, 223, 235, 244, 248, 251, 277, 281, 285, 311, 312, 313. *See also* Cleon, Diodotos, Pericles, Thucydides
Atkins, P.W. 309
Atrahasis 157
Atran, Scott 8
Auden, W. H. 16
Augustine 61, 114, 154, 256
Austen, Jane 32, 259, 289
Averroës 125
Avicenna 125
Azzam, Abdullah Yusuf 201, 273, 275

B

Babyloniaka 157

References

Bacon, Sir Francis v, 16, 17, 27, 28, 66, 67, 120, 127, 129, 134, 136, 176, 182, 188, 292, 297
Ball, John 104
Barrow, John D. 27
Bates, Marston 39
Benkman, Craig 44
Berki, R.N. 277
Berossus 120
Beyer, Bryan E 151, 153, 163, 220
Bible. *See* Hebrew anthology
Bickerton, Derek 109
birds 42, 43, 44, 45, 50
Black, Antony 281
Blake 296
Blomquist, Jeff 309
Bloom, Harold 163
Bloom, Howard 56, 283
Boehm, Christopher 91
Boer, Roland 296
Boyle, Robert 178, 185
Bragg, Melvyn 249
Brahe, Tycho 194
brain ix, x, xiii, 4, 8, 14, 22, 25, 26, 39, 42, 45, 46, 56, 75, 78, 100, 104, 108, 110, 112, 141, 183, 199, 211, 242, 246, 258, 273, 281, 305, 308
Braudel, Fernand 274, 290
Brecht, Bertold 130
Bronowski, Jacob 66
Brown, Dan 253
Bruno, Giordano 172, 176
Bruns, Gerald 163
buildup/breakdown. *See* cycles
Bunyan, John 229
Bürgi, Jost 194
Burke, Kenneth 95, 163
Burling, Robbins 77
Burns, Robert 10
Burroughs, John 23, 42, 43, 60, 64, 133, 212, 213, 320
Bush, George W. 239, 256, 260, 309

C

Cambridge Platonists 122, 186
capitalism 189, 193, 244, 296, 298, 299. *See also* Smith, Adam
Carlyle, Thomas 122, 124
Carnap, Rudolf 130
Carroll, Sean xi, 3, 309
Cassirer, Ernst 116, 124, 125, 126, 127, 157
Castiglione, Baldassare 291
casual talk 29, 30
chaos ix, 12, 36, 48, 53, 54, 55, 59, 166, 173, 211, 269, 270, 303. *See also* entropy, randomness
chaos theory 173
Chaucer 291
Chayefsky, Paddy 161
Cheney, Donald 239, 260, 309
Chomsky, Noam xiii, 75
Cicero 102
Clausewitz, Carl von xv
Cleon 311, 312

Coleridge 43
colonialism. *See* imperialism
Communism 84, 253, 284, 298
Comrie, Bernard 75
Conner, W. Robert 312
cooperation xv, 39, 74, 76, 77, 82, 96, 157, 287. *See also* groupthink
Copernicus 28, 194, 197, 285
Corballis, Michael 79
Cornford, Francis 140
Cosmides, Leda 75
Cromwell, Oliver 231, 233, 272, 284
cuneiform inscriptions. *See* writing
cycles 60, 61, 62, 123, 210, 214, 245, 267, 287, 303, 317

D

Dalley, Stephanie 18, 272
Darwin, Charles xiv, xviii, 9, 23, 32, 37, 39, 44, 58, 67, 175, 190, 271, 273, 297, 308, 318
Darwin, Erasmus 37
Dawkins, Richard 66, 81
Deacon, Terence 76, 77
Defoe, Daniel 291
Dekker, Thomas 291
Delitzsch, Franz 155, 156, 198, 199, 206, 208
Derrida, Jacques 102, 139
Descartes, René xi, 130, 136, 175, 183, 186, 194

Dewey, Thomas 39, 297, 309
Diamond, Jared 77, 96, 264
Diodotos 311, 312, 313
discourse ix, 25, 111, 114, 281, 325, 331. *See also* dissemination, language development, nomenclature, propaganda, rhetoric
disorder. *See* chaos, turbulence
Dissanayake, Ellen 19
dissemination xi, 252, 253, 255, 259, 289, 306, 310. *See also* discourse, propaganda, rhetoric doctrine, indoctrination, propaganda, rhetoric
Donald, Merlin 77, 111
Donne, John 62, 174, 175

E

Edelman, Gerald M., 129
Edgar, Blake 88
ego. *See* subjectivity
Eicher, David J. 214, 309
Einstein, Alfred xii, 20, 21, 110, 185
Eisenstadt, S. N. 113, 125
Eisenstein, Sergei 268
Eldredge, Niles 44
Eliade, Mircea 157, 246
Eliot, T. S 197
Emerson, Ralph Waldo 122, 184
empires. *See* imperialism

empiricism 23, 51, 68, 96, 122, 130, 134, 136, 176, 182, 187, 193, 194, 195, 196, 210, 250, 301, 309. *See also* Bacon
Engels, Friedrich 59, 91, 129, 136, 180, 230
Englefield, Ronald 95
entropy xi, xii, 5, 59, 129, 269
epic 18, 101, 139, 157, 165, 201, 329, 336
Epic of Creation 18, 101, 157
Erra and Ishum 18, 101, 157
Espinas, Alfred V. 244
essence. *See also* Plato
essences 22, 61, 62, 64, 133, 240
Euripides 223, 315
Eusebius 67, 68
evidence selection x, 59, 129, 309, 313. *See also* fact denial, filtering

F

Ferdinand of Aragon 181
Ferguson, Adam xv, 31, 52, 53, 54, 75, 91
Ferrill, Arther 209
Feyerabend, Paul 29
fiction. *See* poesis
Finklestein, Israel 231
Fischer, David Hackett 33, 52
Fisch, Harold 163
Folger, Tim xiv
Fortunoff, David 130
Foucault, Michel 29, 129
Franklin, Benjamin 23, 40, 264
Frazer, James 116
frontier 251, 263, 275, 276, 281
Frye, Northrop 58, 104, 116, 156, 163
Funkenstein, Amos 147
Furst, Alan 291, 292

G

Galbraith, James K. 261
Galileo Galilei 28, 47, 127, 130, 134, 176, 185, 194, 196, 197, 198, 285
Gamow, George (Georgiy Antonovich Gamov) 13, 56, 174
Gassendi, Pierre 186
Gat, Azar 220
genetics 37, 80
Gessner, Konrad 194
Gibbon, Edward 231, 235
Gilgamesh 151, 157
Gilovich, Thomas 141
Gleick, James 173, 269
Godfrey-Smith, Peter xi, 85, 130, 308
Goffman, Erving xiii, 86
Goldberg, Dave 309
Goleman, Daniel 316, 317
Gomez, Jose Maria viii
Gore, Al 317
Gott, Richard J. 172, 309

Gottwald, Norman K. 257, 258
Gould, Stephen Jay 44, 80
Graves, Robert 123, 223
Gray, Russell 75
Greek 10,000 283, 286, 287, 288
Greenberg, Joseph H. 75
Gribbin, John 130, 309
groupthink ix, 92, 156, 207, 218, 227, 230, 243, 245, 250, 257, 267, 278, 304. *See also* ideology, indoctrination, rhetoric, social orders, war
Gunkel, Hermann 155, 156, 198, 199, 200, 206

H

Hacataeus 116
Hagger, Nicholas 52
Haklyut, Richard 228, 230, 247, 248, 262
Hall, Edward 129
Hallpike, C. R. 125, 206, 209
Hamblin, William J. 203, 205
Hamilton, Alexander 278, 282, 283
Hammurabi Code 140, 163, 255, 273
Harris, Marvin 195
Harris, Stephen L. 269
Hauser, Marc 83
Hebrew anthology 69, 113, 116, 117, 119, 149, 151, 152, 155, 198, 202, 210, 230, 255, 273
Hegel, Georg Wilhelm Friedrich xviii, 129, 296, 297
Heidegger, Martin 173
Heraclitus 50, 242
Herbert, George 120
Herodotus 114, 116, 153, 225, 226
Hesiod 69, 107, 114, 116, 124, 128, 160, 272
hierarchy 55, 83, 91, 95, 102, 104, 148, 149, 182, 219, 290
Hill, Reginald 288
historiography xv, 191, 240, 289. *See also* Thucydides
Hitler, Adolf 88, 92, 93, 207, 223, 242, 268, 276
Hobbes xviii, 51, 87, 125, 136, 141, 143, 144, 147, 162, 175, 176, 178, 179, 180, 182, 183
Holinshed 129
Homer 18, 38, 107, 116, 124, 152, 158, 160, 164, 229, 234, 235, 315
Hooke, Robert 14, 15, 45
Horace 143, 234
House of Atreus 164, 223
Hoyle, Fred 321
Hubble, Edwin 196
Humboldt, Alexander von 14, 190, 196, 197, 297
Hume, David 136, 141, 175, 183, 186, 194, 231
Hurford, James R. 99
Huxley, Thomas 16, 32, 33, 190, 191

I

ideology 35, 129, 154, 229, 238, 240, 241, 242, 244, 256, 276, 278, 288, 295, 297, 299
immigrants 41, 42, 237, 250, 263, 275, 289
imperialism 151, 152, 360. *See also* war
indoctrination 29, 90, 95, 146, 147, 156, 185, 219, 225, 240, 252, 253, 261, 267, 271, 284. *See also* propaganda, Rhetoric irregularity; *See* chaos, turbulence

J

Jackendoff, Ray 75, 100
James, Henry 16
James, William 297
Janke, Paul R. 306
Jaspers, Karl 125
Jeans, Sir James 15
Jefferson, Thomas xv, 264, 292, 293
Johnson, Dominic 317
Johnson, Samuel 261
Joyce, James 171, 172
Jung, Carl 157

K

Kafka, Franz 121, 191
Kahneman, Daniel 272
Kant, Immanuel 22, 129, 296
Karp, Walter 238, 239
Kaufman, Stuart 13, 59, 271
Keeley, Lawrence H. 205, 210, 220
Keneally, Thomas 243
Kepler, Johannes 194
Klein, Richard 88
Klinkenborg, Verlyn 57
Knightley, Phillip 276
Kolakowski, Leszek 129
Kuhn, Thomas xi

L

Lamarck, Jean-Baptiste 37, 346
Lang, Andrew 111
Lang, Bernard 128
Langer, Susanne 116
Langland, William 291
language development 74, 76. *See also* discourse
le Carré, John 291
Lecourt, Dominique 193
Leeming, David 111
Leeuwenhoek, Antonie van xvii, 194
Leibniz 51, 133, 147, 175, 186, 187, 194
Lemaître, Georges 196
Lemche, Niels Peter 117, 142, 143

Leopold, Aldo 24, 65, 128, 212
Lily 102
limbo xi, 59, 121, 319, 320
literary kinds 147, 151
Livy 234
Locke, John 128, 136, 141, 143, 144, 162, 175, 178, 183, 184, 185, 186, 187, 194, 200
Loeb, Avi 303
Logan, Robert K. 88, 109
Lucan xiv, 38, 158, 160, 161, 164
Lucretius 124, 229, 230, 293
Lukác, György 129
Lyell, Charles 37, 308

M

MacArthur, Douglas 272
Machiavelli 181, 231
Macpherson, Jay 156
Madison, James xv
Magritte, René 167
Malory, Thomas 126, 158, 159
Malthus, Thomas 39, 166, 175, 194, 195, 228, 273, 293, 318
Mandelbrot sets 27, 173
Marvell, Andrew 22, 186, 199
Marxism 244, 296, 299. *See also* ideology, indoctrination
Marx, Karl 59, 129, 180, 296, 297, 298
Mateen, Omar 260
Mayr, Ernst 177
McCutcheon, John 250

Melian dialogues 236, 311
Mesha Stele (Moabite Stone) 201, 206
Messiah 141, 142, 145, 162, 233
Meyers, Carol 222, 232, 252
Michelangelo 163
Migne, J. P. 119
migration. *See* immigrants
militancy. *See* war
Miller, Wiley (Non Sequitur) xvii
Milton, John xviii, 54, 55, 84, 85, 86, 102, 139, 140, 141, 143, 147, 148, 150, 152, 154, 158, 159, 160, 161, 162, 164, 166, 167, 180, 184, 186, 205, 221
mind. *See* brain
Mithen 98
Mithen, Steven 98
Mlodinow, Leonard 309
mode of production 81, 137, 195, 217, 218, 251, 264, 285
modernism 171, 175, 193, 195. *See also* postmodernism
Montaigne 23
Montesquieu 51, 52
moral philosophy 57, 68, 118, 122, 127, 129, 197, 219, 235, 304
More, Thomas 181, 284
Morgan, Lewis H. 91, 136, 180, 293
Muir, John 37, 38, 39, 40
Mytileneans 311, 312, 313

N

Napier, John 194
national identity 217, 243, 245, 264, 314. *See also* indoctrination, propaganda, rhetoric, turbulence, war
natural continuum viii, ix, xii, 4, 187, 214, 270, 298, 308. *See also* natural history
natural heterogeneity 10, 37
natural history v, viii, ix, xii, xiii, xvi, xvii, xviii, 8, 11, 12, 13, 16, 20, 21, 25, 26, 37, 39, 41, 49, 51, 55, 57, 68, 85, 114, 115, 123, 124, 127, 129, 130, 134, 136, 147, 175, 176, 177, 181, 182, 184, 189, 191, 194, 196, 199, 210, 211, 230, 235, 240, 247, 253, 270, 295, 297, 304, 306, 309, 310. *See also* natural continuum
naturalism 5, 7, 10, 46, 134, 191, 314, 318
naturalist essay 23
natural philosophy 15, 21, 23, 26, 44, 49, 51, 57, 97, 122, 127, 157, 182, 196, 197, 267, 304, 308, 318, 320
Newton, Sir Isaac 6, 8, 48, 110, 194
Nicias 314, 315, 342
Niditch, Susan 220
Nisbet, Robert 296
nomenclature 7, 11, 20, 26, 27, 32, 33, 35, 36, 37, 46, 48, 51, 54, 55, 56, 62, 114, 132, 240
Norris, Margot 276

O

objectivity ix, xi, xii, xvi, 20, 40, 98, 157, 186, 302. *See also* subjectivity
Old Testament. *See* Hebrew anthology
omni terminology. *See* nomenclature
oracular sermon 101, 139, 140, 141, 144, 149, 151, 152, 153, 155, 175, 211, 217
Osama bin Laden 201, 275
Ovid 7, 10, 55, 61, 124, 202

P

Pagels, Elaine viii, 166
Parmenides 33, 50, 301
parousia 145, 153. *See also* Messiah
Parr, Mr. 310
patterns 7, 8, 31, 58, 62, 75, 92, 116, 191, 227, 288, 304
Pears, Iain 30
Peattie, Donald Culross 41, 211, 213, 317, 318, 319, 320
Penrose, Roger 26, 309
Pericles 222, 238, 281, 311, 313

Periodic Table of Elements 4, 27, 33, 171, 307
Peter 143, 144
Peterson, Dale 80
Petronius vii
Philo 63, 114, 115, 117, 118, 119, 122, 123, 175, 198, 304
Pinker, Steven 75
Plato 22, 23, 61, 62, 63, 102, 107, 114, 115, 119, 124, 129, 130, 131, 132, 133, 134, 175, 188, 238, 301
playing dead 84, 86
Plutarch xvi, 290, 291, 314, 315
poesis viii, xviii, 9, 16, 18, 20, 57, 68, 81, 100, 101, 126, 161, 182, 196, 207
Polybius 30, 38
Pope, Alexander 103, 190
Popper, Karl 29, 130
postmodernism 11, 171, 173, 193, 195, 241, 269, 297, 301, 302. *See also* Modernism
Power, Henry 14, 45
propaganda viii, ix, xiii, 31, 84, 92, 93, 188, 207, 208, 223, 225, 232, 243, 259, 276, 286, 287. *See also* indoctrination
prosody 98, 99, 100, 101, 102, 103, 104, 156
Protagoras 315
Puritans 104, 141, 145, 176, 219, 224, 256, 272
Pythagoras 175

Q

Quine, Willard 130

R

Raglan, Lord 157, 246
Rajneeshees 201
randomness 54, 58, 173, 211
reified abstraction. *See* essences, Plato
relativity 20, 21, 41, 174, 185, 186, 187, 190
reliability 25
religion 29, 50, 111, 155, 175, 184, 197, 199, 226, 231, 262, 275, 282, 314. *See also* Hebrew anthology
representation xix, 26, 27, 29, 43, 68, 114, 282, 305, 309. *See also* dissemination
rhetoric x, 19, 89, 95, 100, 101, 102, 104, 114, 142, 145, 147, 185, 198, 201, 205, 211, 225, 226, 227, 229, 238, 239, 240, 264, 271, 281, 282. *See also* discourse, indoctrination, propaganda
Ridley, B. K. 27
ritual 19, 47, 48, 49, 50, 96, 123, 151, 154, 156, 157, 158, 160, 161, 198, 204, 250, 255, 277
Roaf, Michael 254

Rorty, Richard 297, 302
Rousseau, Jean-Jacques xviii, 59, 91, 125, 136, 180, 230
Roux, Georges 319
Royal Society 14, 23, 37, 136, 176, 177, 186, 194, 297
Ruskin, John 13, 59
Rutherford, Ernst 11
Ryder, Brett 305
Ryle, Gilbert 111, 116

S

Sagan, Carl 125, 189, 310
Sarles, Harvey 76
Scaruffi, Piero 276
Schatzman, Evry 193
Schmidt, Alfred 297
Schneidau, Herbert 163
Searle, John R. 301, 302
semantics. *See* discourse
Seneca 102
Service, Elman R. 180, 204, 301
Seters, John van 202
Shakespeare 62, 63, 159, 181, 182, 243, 257, 261, 273, 283
Shermer, Michael 141
Sidney, Sir Philip xviii, 17, 18, 102
Silberman, Neil Asher 231
Smith, Adam 129, 294, 295, 296, 297, 298. *See also* capitalism
Smith, John Maynard 80
Smolin, Lee 29

Social orders 34, 95, 96, 136, 173, 244, 245, 270, 271, 273, 286, 287, 301. *See also* Groupthink
Song of Roland 158
Sophocles 315
Sorokin, P. A. 289
Spengler, Oswald 246
Spenser, Edmund 126, 127, 158, 159, 228, 229
Spinoza, Baruch xviii, 130, 175, 186, 284
Sprat, Thomas 23, 176
Statius 18, 158, 283
Sternberg, Meir 163
Stevens, Wallace 20, 97, 197
Stewart, Ian 8, 48, 130, 307, 309
subjectivity 6, 20, 310. *See also* objectivity
supreme fiction 97, 197, 238
Swift, Jonathan 175, 187, 188, 189, 194
Systematic Nomenclature. *See* discourse, nomenclature
Szathmàry, Eörs 80

T

Tacitus 225
terrorism 4, 5, 197, 256, 275
Thales 125
Third Reich 31, 32, 154, 242, 276
Thompson, Thomas L. 125
Thucydides 107, 108, 116, 124, 128, 225, 229, 230, 235, 236,

237, 238, 239, 240, 251, 252, 293, 314
Tocqueville, Alexis de xiv
Tolstoy, Leo 191
Tooby, John 75
topography 7, 36, 37, 41, 147, 177, 211, 284
Traherne, Thomas 122
Trigger, Bruce G. 126, 162, 221
Trump, Donald 88
Tudge, Colin 66, 92
Turbulence 153, 268, 270, 273, 287, 289, 290, 291, 294, 296. *See also* chaos, war
Turing, Alan 26
Turner, Frederick Jackson 263
Twain, Mark vi, 100, 146, 151, 205
Tyson, Neil deGrasse 310

U

Uffenheimer, Benjamin 112
Usener, Hermann 126

V

Vaughan, Henry 121, 122
Verdi, Giuseppe 209
Vico 147
Virgil 18, 38, 140, 152, 154, 158, 160, 221, 234

W

Wallace, Alfred xiv, 9
Walton, Sir Isaac 37
war viii, xiv, xv, xvi, 3, 6, 51, 52, 53, 54, 89, 92, 93, 101, 112, 116, 119, 141, 150, 151, 153, 167, 180, 181, 200, 205, 206, 208, 209, 217, 218, 220, 221, 224, 225, 226, 229, 235, 237, 238, 242, 251, 252, 259, 261, 271, 273, 275, 280, 285, 289, 309, 312, 314, 315. *See also* epic, imperialism, oracular sermon, propaganda
Wellhausen, Julius 142, 202
Wenke, Robert J. xviii
Widengren, Geo 257
Wilson, Edward O. 54, 320
Wilson, John A. 111
Wilson, Robert R. 142
Wittgenstein, Ludwig 173
Wordsworth, William 64, 122
Wrangham, Richard 80
writing x, 23, 37, 103, 109, 111, 164, 180, 184, 197, 202, 206, 209, 210, 211, 223, 225, 244, 254, 255, 296, 304

X

Xenophon 127, 131, 238, 283, 286, 287, 361
Xerxes 49, 223, 225, 226, 286, 363

Y

Yonge, C. D. 119

Z

Zagorin, Perez 255, 289
Zeno 33, 50, 114, 301
Zimbardo, Philip 90
Zimmer, Carl 68
Zoroaster 120